Frost Action in Soils

Fundamentals and Mitigation in a Changing Climate

Edited by
Sally A. Shoop, Ph.D., P.E.

Prepared by the Frozen Ground Committee and the Transportation and Infrastructure Committee of the Cold Regions Engineering Division of the American Society of Civil Engineers

ASCE

Published by the American Society of Civil Engineers

Library of Congress Cataloging-in-Publication Data

Names: Shoop, Sally A. (Sally Annette), editor. | American Society of Civil Engineers. Cold Regions Engineering Division.
Title: Frost action in soils : fundamentals and mitigation in a changing climate / prepared by the Frozen Ground Committee and the Transportation and Infrastructure Committee of the Cold Regions Engineering Division of the American Society of Civil Engineers ; edited by Sally A. Shoop, P.E., Ph.D.
Description: Reston : American Society of Civil Engineers, 2020. | Includes bibliographical references and index. | Summary: "Frost Action in Soils: Fundamentals and Mitigation in a Changing Climate reviews and updates the state of knowledge on frost-action fundamentals, the impact of climate change, and mitigation of frost action on pavements and other structures"– Provided by publisher.
Identifiers: LCCN 2019050270 | ISBN 9780784415085 (paperback) | ISBN 9780784481509 (pdf)
Subjects: LCSH: Frost heaving. | Frozen ground. | Soil freezing.
Classification: LCC TA713 .F759 2020 | DDC 624.1/5136–dc23
LC record available at https://lccn.loc.gov/2019050270

Published by American Society of Civil Engineers
1801 Alexander Bell Drive
Reston, Virginia 20191-4382
www.asce.org/bookstore | ascelibrary.org

Errata: Errata, if any, can be found at https://doi.org/10.1061/9780784415085.

ISBN 978-0-7844-1508-5 (print)
ISBN 978-0-7844-8150-9 (PDF)

Manufactured in the United States of America.
26 25 24 23 22 21 20 1 2 3 4 5

Contents

Preface

Significant technological advancements addressing frost action in soils have occurred in the years since the 1984 ASCE Frost Action Monograph. In addition, climate change poses questions regarding associated effects on freeze–thaw action. This publication serves to update the current state of the knowledge on frost action for cold regions engineering practitioners. The first section presents the fundamentals of frost heave and thaw weakening, effects on roads and other structures, and the projected effects of climate change on frost action. The second section presents mitigation of frost heave and thaw weakening within pavement structures. The manuscript concludes with three case studies dealing with frost action and mitigation for buildings, roadways, and airfields. The intent of this publication is to describe the challenges of cold regions engineering in a changing climate and to provide state of the art tools for addressing these.

The ASCE Cold Regions Engineering Division (CRED) is extremely grateful to all of the authors and reviewers for their volunteer efforts. The manuscript was initiated and developed by the CRED Frozen Ground Committee and the Transportation and Infrastructure committee and was approved by the CRED Executive Committee. Each of the chapters has been reviewed for technical accuracy and understanding. The authors are included at each chapter heading and listed on the following page, along with the reviewers. The manuscript also benefited from the efforts of our extremely competent associate editor Wendy Wieder of CRREL and technical copy editor Emily Moynihan of BlytheVisual, LLC, NH.

My sincere respect and appreciation to those dear friends and colleagues who contributed and persevered throughout the publication process.

Sally A. Shoop, Ph.D., P.E., M.ASCE
Editor
Chair, ASCE Cold Regions Engineering Division Executive Committee
Former Chair, Frozen Ground Committee
Research Civil Engineer, US Army Engineering Research
and Development Center,
Cold Regions Research and Engineering Laboratory (ERDC-CRREL)

Contributing Authors and Reviewers

Contributing Authors

Jean-Pascal Bilodeau, Ph.D., Laval University
Kevin Bjella, P.E., CRREL
Gilles Brillant, Québec Ministry of Transportation
Kuo-Chieh Chao, Ph.D., P.E., M.ASCE, Engineering Analytics, Inc.
Billy Connor, P.E., MASCE, University of Alaska Fairbanks
Guy Doré, Ph.D., Laval University
Jeremiah Drage, P.E., Golder Associates
Ariel Ellison, CRREL
W. Spencer Guthrie, Ph.D., P.E., M.ASCE, Brigham Young University
Karen S. Henry, Ph.D., P.E., M.ASCE, US Air Force Academy (Ret.)
Travis Logan, Ph.D., Laval University
Jon Maakestad, CRREL
John Nelson, Ph.D., P.E., D.GE, F.ASCE, Engineering Analytics, Inc.
Antti Nurmikolu, Ph.D., Tampere University of Technology
Daniel Overton, P.E., F.ASCE, Engineering Analytics, Inc.
Sally A. Shoop, Ph.D., P.E., M.ASCE, CRREL
Denis St-Laurent, Québec Ministry of Transportation

Technical Reviewers

Heather Brooks, Ph.D., P.E., M.ASCE, BGC Engineering Inc.
Robert A. Eaton, P.E., M.ASCE, FROST Associates, CRREL (Ret.)
Karen S. Henry, Ph.D., P.E., M.ASCE, US Air Force Academy (Ret.)
V. R. (Sivan) Parameswaran, Ph.D., P.Eng., M.ASCE, National Research Council of Canada (Ret.)
Sally A. Shoop, Ph.D., P.E., M.ASCE, CRREL
Wendy L. Wieder, Ph.D., P.E., CRREL

Associate Editor

Wendy L. Wieder, Ph.D., P.E., CRREL

Technical Copy Editor

Emily Moynihan, BlytheVisual, LLC

Acronyms and Abbreviations

AAAF	Allen Army Air Field
AC	asphalt concrete
AFI	air freezing Index
AR5	IPCC's fifth assessment report
BRRC	Belgium Road Research Laboratory
CBR	California bearing ratio
CMIP5	World Climate Research Programme's Coupled Model Intercomparison Project Phase 5
CRED	Cold Regions Engineering Division
CRREL	Cold Regions Research Engineering Laboratory
DOT	Department of Transportation
EICM	enhanced integrated climatic model
FERF	Frost Effects Research Facility
FGA	Fort Greely, Alaska
FI	freezing index
FTC	freeze–thaw cycles
FWD	falling weight deflectometer
GMFD	global meteorological forcing dataset
GPR	ground-penetrating radar
GSSI	Geophysical Survey Systems Incorporated
HMAC	hot mix asphalt concrete
IFSTTAR	Institut français des sciences et technologies des transports, de l'aménagement et des réseaux
IPCC	Intergovernmental Panel on Climate Change
IRI	international roughness index
j.d.	Julian day
LCPC	Laboratoire central des ponts et chaussées
LL	liquid limit
LSFP	limited subgrade frost penetration
MAST	mean annual surface temperature
MEPDG	mechanistic-empirical pavement design guide
MTO	Ontario Ministry of Transportation
NASA	National Aeronautics and Space Administration
NCHRP	National Cooperative Highway Research Program
NEX-GDDP	NASA Earth Exchange Global Daily Downscaled Projections
NOAA	National Oceanic and Atmospheric Administration
OPAC	Ontario Pavement Analysis of Cost
PI	plasticity index

PSI	pavement serviceability index
RCI	ride comfort index
RCP	representative concentration pathway
RSS	reduced subgrade strength
SC-SM	silty clayey sand
SMDC	US Army Space and Missile Defense Command
SP	segregation potential
SWE	snow water equivalent
TCCRE	Technical Council on Cold Regions Engineering
TRRL	Transport and Road Research Laboratory (UK)

PART I

Fundamentals and Climate Context

CHAPTER 1

Introduction and History

Billy Connor
Sally Shoop

1.1 INTRODUCTION

Frost heave is a deformation of soil caused by the freezing of soil pore water and migration of water to the freezing front. The main direction of the deformation is in the direction of heat loss. Because frost action can cause catastrophic damage to a variety of infrastructure elements in cold regions, it is important to understand its mechanics. The high cost of repair of damaged facilities warrants detailed studies of frost action by cold regions researchers and engineers seeking to better understand the process and how to mitigate frost and thaw impacts on infrastructure (Cary et al. 1979; Doré and Konrad 1999; Konrad 1987, 1989; Konrad and Morgenstern 1982; Doré et al. 1996; Penner 1959, Henry 2000, Guthrie et al. 2016).

1.2 HISTORY

In 1938, J.O. Osterberg translated the works of Dr. Gunnar Beskow into a treatise that contained much of the history and knowledge of frost heave up to 1935. In 1991, Patrick Black and Mark Hardenberg combined Beskow's works with the early works of Stephen Taber, creating Special Report 91-23, published by the Cold Regions Research Engineering Laboratory (CRREL). Following is a brief history, as described in "Soil Freezing and Frost Heaving with Special Applications to Roads and Railroads" (contained in Black and Hardenberg 1991).

Frost heave has been recognized since at least the late 1600s. Beskow explains that the early literature contains terms such as "frost boil" and "pipkrake"; Urban Hjärne (1694) uses the latter in 1694, and describes "freezing up of stones from the ground" along with "earth heaving" and "earth shooting." This phenomenon is more completely explained in 1706. By the middle of the 1700s, frost heaving had been associated with ground freezing, when E.O. Runeberg (1765) stated, "At the depth where the frozen soil stops, I found ice granules, lumps of clear blue ice in big pieces." Tilas observed in the late 1700s that frost heave had done damage to culverts and bridge piers. At the time, thaw weakening was not recognized as a cause of damage because its effects were overshadowed by damage caused by flooding, snowmelt, and thawing of ice.

In 1914, Simon Johnson found that water flows to the freezing layers during freezing, thereby increasing the volume of water in the soil. In 1929, Taber completely discredited the general assumption that frost heave is caused by the expansion of freezing water; he argued that it is caused by the growth of ice lenses in soil. Taber's experiments replaced water with benzene and nitrobenzene, which decrease in volume when frozen; however, during the freezing process, the soil continued to heave as long as the soil column had access to the liquid at the bottom.

The earliest theoretical work of note was that of physicist G.J. Bouyoucos (1920) of Michigan State University. Bouyoucos discovered that not all water in soil freezes at the same temperature.

Sweden was the first country to seriously consider the problem of frost heave. The Swedish Institute of Roads held a conference in Luleå in October 1925, and afterward, the Swedish Geological Survey and the Swedish Institute of Roads jointly began a "theoretical-practical" investigation of the frost problem. Work began in earnest in 1927, resulting in a number of papers published between 1928 and 1933.

The United States, Soviet Russia, Norway, and Germany began their own investigations during this same time. The US frost investigations began with some experiments in 1926 by Stephen Taber and A. Casagrande. This work was expanded by Popke (see Black and Hardenberg, 1991), who suggested that water is transported by the flow of water vapor, which condenses on the soil particles as they freeze. Others, including Gustaf Dahlberg, argued that this is essentially impossible, because the heaving soil is always filled with capillary water, and that even if it were possible, diffusion would be so slow as to be inconsequential. In 1929, Taber's careful experiments showed that water may be moved to the freezing front through vapor transport. However, this is not the only transport mechanism; and Taber agreed with Johansson (1914) that capillary transport is an important process.

In 1931, Casagrande observed a correlation between the percentage of soil particles finer than 0.002 mm and frost heave. However, Casagrande never intended the field observations of a few soils to become a specification for frost susceptibility.

A few years later, R.K. Schofield (1935) proposed the pF scale, which described the relationship between the free energy of water in moist soil and the freezing-point depression. Others believed Schofield's work to be superficial and proposed using thermodynamics to understand the freezing of soil moisture. Soil physicists who pursued Schofield's pF scale ultimately abandoned soil freezing studies by the mid-1940s.

The US Arctic Construction and Facilities Engineering Laboratories accumulated the results of systematic empirical frost-heave tests during the 1950s, using equipment similar to that used by Taber. However, they were unable to develop a relationship between frost heave and engineering properties that was any better than that proposed by Casagrande in 1931.

In the late 1950s, researchers developed theories to explain the frost-heave process. K. Jackson and B. Chalmers (1958) proposed that heaving was an irreversible process dependent on the freezing of supercooled water. L. Gold (1957) proposed a frost-heave model based on the surface tension of the ice–water interface. R.D. Miller of Cornell University suggested an osmotic model based on a series of experimental observations and theories of frost-heave mechanisms (Cass and Miller 1959). All of these were based on the lens growth proposed by Taber (Black and Hardenberg 1991).

Throughout the 1960s, Miller and his students continued to explore the theoretical basis for frost heave. They suggested that the Clapeyron equation was confirmed by an osmotic model of the unfrozen water film with a surface tension model at the base of the ice lens.

Konrad and Morgenstern (1980, 1981) extended the work of Miller by combining unfrozen pore water suction, calculated by the Clausius-Clapeyron equation at the base of the ice lens, and thermodynamic equilibrium at the same location. They determined that the water uptake into the frozen fringe (the area between the frost front and the pore water freezing point) is proportional to the temperature gradient within the frozen fringe. The constant of proportionality, which Konrad and Morgenstern termed the *segregation potential*, is a function of the segregation freezing temperature, the temperature at the base of the ice lens, and the permeability of the frozen fringe. Konrad continued this work by researching changes in effective void ratio due to freezing and thawing of clayey silt soils and the influence of heat extraction on frost heave (Konrad 1987, 1989).

Péwé and Paige (1963) published a paper describing frost heave on the Alaska Railroad. This is one of the earliest investigations of the impact of frost heave on public infrastructure. Péwé recorded as much as 14 in. per year frost jacking on bridge pilings, requiring corrective action by the railroad company. He suggested a detailed explanation of the mechanics of frost heaving and frost jacking. Péwé and Paige's paper provides an excellent application of theory to an engineering problem, establishing an engineering practice still followed.

In the 1970s and 1980s, frost action became a major research topic for engineering applications, including differential frost heaving, frost jacking, and spring thaw weakening. Research tended to focus on understanding how to select soil gradations to minimize the impacts of frost action. Penner and Crawford (1983) described the relationship between climate and seasonal frost penetration, including the mechanism of frost heaving and frost susceptibility of soils. The report discussed several examples of impacts of frost action on infrastructure. In each case, mitigation techniques provided future engineers with insight into avoiding impacts of frost heaving.

Esch et al. (1980) reported the relationship between the percentage passing the 0.075 mm and the 0.002 mm sieves and pavement structure performance in Alaska. They suggested that these parameters may be the most important indicator of degradation of pavement performance due to thaw weakening. In contrast, the R-value design method adopted from the California Department of Transportation (2008) was of no value in predicting pavement performance in Alaska. This research was typical of that being performed throughout the United States and Canada in the 1980s relating soil properties to thaw weakening.

Esch (1969, 1971) first reported on the use of insulation to control frost heave. This first attempt used expanded polystyrene insulation to reduce frost heaving near Anchorage, Alaska. While the effort was initially successful, the insulation absorbed water, rendering it ineffective as time passed. Later, the Alaska Department of Transportation and Public Facilities successfully used extruded insulation manufactured by Dow Chemical to reduce frost heaving (Esch 1986, 1994). However, in some cases, differential icing occurred on the roadways over the insulation. As a result, Dow refused to sell insulation for use in roadways, citing liability concerns. Research showed that if the insulation was placed about 3 ft below the pavement surface, frost action could be controlled while minimizing differential heaving.

In the 1980s and 1990s, two full-scale experimental research facilities were established to study frost effects on pavement systems: the MnROAD facility operated by the Minnesota Department of Transportation and CRREL's Frost Effects Research Facility. MnROAD is a pavement test-track facility on Interstate 94 about 30 miles northwest of Minneapolis (http://www.dot.state.mn.us/mnroad/). It has over 50 instrumented test sections on both its 3.5 mi interstate and its 2.5 mi low-volume road. Over the last 22 years, the facility has hosted projects researching many different topics—seasonal frost heave; seasonal load limits, including spring load restrictions and winter overloads; low-temperature cracking behavior of asphalt pavements; and cold regions' seasonal effects on pavement layers—for inputs into advanced designs. The data collected at MnROAD is saved in the MnROAD database and available for free on the MnROAD website. Initial construction of the facility was completed in 1994; as test sections fail or the research is completed, new test sections are designed and built around local, state, and regional and national research needs. MnROAD has also partnered with the Minnesota Local Road Research Board, which has funded both the low-volume road operations and many research projects of interest to the local city and county engineers in Minnesota over the life of the project. To expand its capabilities, MnROAD has also partnered with other state departments of transportation, the Federal Highway Administration, universities, industry, associations, and international partners to achieve its research goals for its customers.

Around the same time, the US Army Corps of Engineers constructed the Frost Effects Research Facility (FERF) at the Cold Regions Research and Engineering Laboratory in Hanover, NH. The FERF is a temperature- and groundwater-controlled building designed to generate freezing and thawing of full-scale pavement or other structures year-round, simulating winter as often as

12 times per year, depending on the structure (Eaton 1989). FERF hosted several projects instrumental in the study of frost effects on soils, pavement, and other structures where soil, temperature, and groundwater conditions were controlled. Studies included thermal cracking of pavement (Kanerva et al. 1992); asphalt concrete pavement instrumentation, performance, accelerated testing, and non-destructive testing (Janoo and Berg 1990; Janoo and Cortez 2002; Janoo et al. 2003a, 2003b; Janoo and Cortez 2003; Janoo and Shoop 2004); freeze/thaw and geosynthetics in paved and unpaved structures (Shoop and Henry 1991; Kestler and Nam 1999; Shoop et al. 2003; Perkins and Cortez 2005; Henry et al. 2009); freeze–thaw effects on moisture migration, soil strength, vehicle mobility, and soil erosion (Shoop and Bigl 1997; Shoop 1993; Berliner and Shoop 1992; Gatto 1998, 2000); and ground anchors in warming permafrost (Liu et al. 1999). This is a sampling of the studies where FERF was instrumental in advancing knowledge of freeze–thaw action in soil, pavements and other structures through large-scale controlled experiments.

1.3 CONCLUSION

Based on this early research and subsequent studies and analysis, practitioners now have the tools to minimize or eliminate frost action in roadways, railways, and airport embankments. The chapters of this book describe the basic principles of frost and thaw action and best practices for managing its effects.

References

Berliner, E., and S. Shoop. 1992. "Instrumentation for vehicle mobility testing in the frost effects research facility." In *Road and airport pavement response monitoring systems: CRREL report MP 3071*. V. C. Janoo and R. A. Eaton, eds., 12–26. New York: ASCE.

Black, P. B., and M. J. Hardenberg. 1991. *Historical perspectives in frost heave research: The early works of S. Taber and G. Beskow*. Special Rep. No. 91–23. Hanover, NH: US Cold Regions Research and Engineering Laboratory.

Bouyoucos, C. J. 1920. "Degree of temperature to which soils can be cooled without freezing." *J. Agric. Res.* **20**: 267–269.

Caltrans (California DOT). 2008. Chap. 600 in *Highway design manual*. Sacramento, CA: Caltrans.

Cary, J. W., R. I. Papendick, and G. S. Campbell. 1979. "Water and salt movement in unsaturated frozen soil: Principles and field observation." *Soil Sci. Soc. Am. Proc.* **43** (1): 3–8.

Casagrande, A. 1931. "Discussion of frost heaving." In *Proc., 11th Annual Meeting of the Highway Research Board*. Washington, DC.

Cass, L., and R. D. Miller. 1959. *Role of the electric double layer in the mechanism of frost heaving*. Research Rep. No. 49. Wilmette, IL: US Army Snow, Ice, and Permafrost Research Establishment.

Doré, G., and J.-M. Konrad. 1999. "Deterioration model for pavements in frost conditions." *Trans. Res. Rec.* **1655** (1): 110–117.

Doré, G., J.-M. Konrad, and M. Roy. 1996. "Early research findings on the performance of instrumented pavement sections with different frost protections." In *Proc., 1996 TAC Annual Conf.* Charlottetown, Prince Edward Island, Canada.

Eaton, R. 1989. "Unique new cold weather testing facility." In *Proc., 5th Int. Cold Regions Engineering Specialty Conf.*, St Paul, MN.

Esch, D. C. 1969. *Subgrade insulation for frost heave control construction: Instrumentation and first year study.* Fairbanks, AK: Alaska Dept. of Highways.

Esch, D. C. 1971. *Subgrade insulation for frost heave control: Summary of second and third winter's performance.* Fairbanks, AK: Alaska Dept. of Highways.

Esch, D. C. 1986. *Insulation performance beneath roads and airfields in Alaska.* Fairbanks, AK: Alaska DOT and Public Facilities.

Esch, D. C. 1994. *Long-term evaluations of insulated roads and airfields in Alaska.* Fairbanks, AK: Alaska DOT and Public Facilities.

Esch, D. C., R. L. McHattie, and B. Connor. 1980. *Frost susceptibility ratings and pavement structure performance.* Fairbanks, AK: Alaska DOT and Public Facilities.

Gatto, L. 1998. "Freeze-thaw effects on vehicular ruts and natural rills: Importance to soil-erosion and terrain modelling." In *Proc., Modelling Soil Erosion, Sediment Transport and Closely Related Hydrological Processes*, W. Summer, E. Klaghofer, and W. Zhang, eds., Vienna, 71–79.

Gatto, L. 2000. "Soil freeze-thaw induced changes to a simulated rill: Potential impacts on soil erosion." *Geomorphology* **32** (1–2): 147–160.

Gold, L. H. 1957. "A possible force mechanism associated with the freezing of liquids." *Highway Res. Board Bull.* **168**: 65–72.

Guthrie, S., S. Shoop, and J. Ulring. 2016. "Frost damage in pavement: Causes and cures." *J. Cold Reg. Eng.* **30** (3): 02516001.

Hjärne, U. 1694. *Een kort Anledning till åtskillige Malm- och Bergarters, Mineraliers, Wäxters, och Jordeslags sampt flere sällsamme Tings, effterspöriande och angifwande* [A brief guide to discovering and specifying various types of ores and mountains, minerals, plants, and soils, together with several unusual things] [In Swedish.] Stockholm, Sweden.

Henry, K. 2000. *A review of the thermodynamics of frost heave.* TR-00-16. Hanover, NH: USACE, Engineer Research and Development Center.

Henry, K., J. Clapp, W. Davids, D. Humphrey, and L. Barna. 2009. *Structural improvements of flexible pavements using geosynthetics for base course reinforcement.* ERDC/CRREL TR-09-11. Hanover, NH: USACE, Engineer Research and Development Center.

Jackson, K., and B. Chalmers. 1958. "Freezing liquids in porous media with reference to frost heaving in soils." *J. Appl. Phys.* **29** (8): 1178–1181.

Janoo, V., and R. Berg. 1990. *Predicting the behavior of asphalt concrete pavements in seasonal frost areas using nondestructive techniques.* CRREL CR 90-10. Hanover, NH: USACE, Engineer Research and Development Center. Columbus, Ohio.

Janoo, V., and E. Cortez. 2002. "Some observations of base course performance during accelerated pavement testing." In Vol. 2 of *Proc., 6th Int. Conf. on Bearing Capacity of Roads and Airfields*, A. Gomes Correia and F. E. F. Branco, eds., Lisbon, Portugal, 1247–1254.

Janoo, V., and E. Cortez. 2003. "Development of a subgrade failure criteria using accelerated pavement testing." In Vol. 21 of *Proc., 11th Road Engineering Association of Asia and Australasia on ARRB Transport Research*, Cairns, Australia.

Janoo, V., E. Cortez, L. Irwin, W.-S. Yang, and K. Petros. 2003a. "Influence of subgrade type and moisture content on failure criteria." In Vol. **1** of *Proc., Int. Conf. on Highway Pavement Data, Analysis and Mechanistic Design Applications*, 7–11. Columbus, Ohio.

Janoo, V., L. Irwin, and R. Haehnel. 2003b. *Pavement subgrade performance study: Project overview*. ERDC/CRREL TR-03-5. Hanover, NH: USACE, Engineer Research and Development Center.

Janoo, V., and S. Shoop. 2004. "Influence of spring thaw on pavement rutting." In *Proc., 6th Int. Symp. on Pavements Unbound (UNBAR 6)*, Nottingham, UK, 115–124.

Johansson, S. 1914. *Die Festigkeit der Bodenarten bei verschiedenem Wassergehalt*. S.G.U., Ser. C, nr. 256 (Arsbok 7). Uppsala, Sweden: Geological Survey of Sweden.

Kanerva, H., T. Vinson, A. Brickman, and V. Janoo. 1992. "Low temperature cracking experiment at USACRREL frost effects research facility." In Vol. **37** of *Proc., Canadian Technical Asphalt Association*, 289–307. Victoria, British Columbia.

Kestler, M., and S. Nam. 1999. "Reducing damage to low volume asphalt-surfaced roads and improving local economies: Update on variable tire pressure project." In *Proc., 10th Int. Conf. on Cold Regions Engineering*, J. E. Zufelt, ed., Lincoln, NH, 461–471.

Konrad, J.-M. 1987. "The influence of heat extraction rate in freezing soils." *Cold Reg. Sci. Technol.* **14** (2): 129–137.

Konrad, J.-M. 1989. "Physical processes during freeze-thaw cycles in clayey silts." *Cold Reg. Sci. Technol.* **16** (3): 291–303.

Konrad, J.-M., and N. R. Morgenstern. 1980. "A mechanistic theory of ice lens formation in fine-grained soils." *Can. Geotech. J.* **17** (4): 473–486.

Konrad, J.-M., and N. R. Morgenstern. 1981. "The segregation potential of a freezing soil." *Can. Geotech. J.* **18** (4): 482–491.

Konrad, J.-M., and N. R. Morgenstern. 1982. "Prediction of frost heave in the laboratory during transient freezing." *Can. Geotech. J.* **19** (3): 250–259.

Liu, H., H. Zubeck, and S. Baginski. 1999. "Evaluation of helical piers in frozen ground." In *Proc., 10th Int. Conf. on Cold Regions Engineering*, J. E. Zufelt, ed., Lincoln, NH, 232–242.

Penner, E. 1959. "The mechanism of frost heaving in soils." *Highway Res. Bull.* **225** (1): 1–13.

Penner, E., and C. B. Crawford. 1983. *Frost action and foundations*. Ottawa: National Research Council Canada.

Perkins, S., and E. Cortez. 2005. "Evaluation of base-reinforced pavements using a heavy vehicle simulator." *Geosynthetics Int.* **12** (2): 86–98.

Péwé, T. L., and R. A. Paige. 1963. *Frost heaving of piles with an example from Fairbanks, Alaska*. Washington, DC: US Government Printing Office.

Runeberg, E. O. 1765. "Anmärkningar om någre förändringar på jord-ytan I allmänhet, och under de kalla Climat I synnerhet." *Kunglica Vetensk Akademiens Handlingar* 81–115.

Schofield, R. K. 1935. "The pF of water in soil." In Vol. 2 of *Proc., Transactions of 3rd Int. Congress on Soil Science*, 37–38. Oxford, United Kingdom.

Shoop, S. A. 1993. "Thawing soil strength measurements for predicting vehicle performance." *J. Terramech.* **30** (6): 405–418.

Shoop, S. A., and S. R. Bigl. 1997. "Moisture migration during freeze and thaw of unsaturated soils: Modelling and large scale experiments." *Cold Reg. Sci. Technol.* **25** (1): 33–45.

Shoop, S. A., and K. S. Henry. 1991. "Effect of a geotextile on water migration and frost heave in a large-scale test basin." *Transp. Res. Rec.* **1307**: 309–318.

Shoop, S. A., M. K. Kestler, J. A. Stark, C. Ryerson, and R. Affleck. 2003. "Rapid stabilization of thawing soils for enhanced vehicle mobility: Field experience and application." *J. Terramech.* **39** (4): 181–194.

Taber, S. 1929. "Frost heaving." *J. Geol.* **37** (5): 428–461.

CHAPTER 2

Mechanisms of Frost Heave and Thaw Weakening

W. Spencer Guthrie
Antti Nurmikolu

2.1 INTRODUCTION

Understanding the mechanisms of frost heave and thaw weakening of soil provides a basis for engineering designs to mitigate these issues when construction in cold regions requires building infrastructure on frost-susceptible soils. Frost heave is manifest as deformation or displacement of soil caused by the migration and freezing of soil pore water. Three conditions are required for frost heave: sustained freezing temperatures, available water, and frost-susceptible soils (Kaplar 1970, Mitchell 1993). Because of spatial variability in these three conditions, differential heaving, rather than uniform heaving, is common in many locations. Differential heaving can damage foundations, buildings, utilities, pavements, railways, and other infrastructure.

In regions subject to seasonal warming, thawing of frozen soil begins at the ground surface as the air temperature begins to rise. In the early stages of thawing, when drainage of meltwater from upper soil layers is often prevented by the underlying frozen soil, the soil can become supersaturated and experience marked reductions in bearing capacity, referred to as thaw weakening (Doré and Zubeck 2009, Guthrie and Zhan 2002, Janoo and Berg 1990). In this condition, the affected soil may not adequately support design loads, and additional damage to infrastructure may occur.

The degree of both frost heave and thaw weakening depends on the amount and distribution of ice that forms in the soil (Taber 1931). Ice can form as a result of both in situ freezing and segregational freezing. In situ freezing is the in-place freezing of water within the pore spaces between soil particles. When soil freezes quickly, in situ freezing is the predominant mechanism of ice formation (Konrad 1984). Typically nucleating near the center of the pore space and gradually expanding outward, ice substantially fills the pores. Although water expands approximately 9% in volume when it freezes, the amount of frost heave resulting from in situ freezing is usually less than about 3% of the depth of the frozen zone in unsaturated soils (McCarthy 1998). The corresponding heaving pressure is also relatively small (Perfect et al. 1988).

In contrast, segregational freezing can generate very significant vertical displacements and heaving pressures, depending on the counteracting overburden pressures (Fowler 1989), and is of primary concern with respect to infrastructure damage. Segregational freezing, which is the freezing of incoming water that moves into the freezing zone from unfrozen soil, occurs when frost-susceptible soils freeze comparatively slowly. When water is available, generally being drawn toward the freezing front from deeper soils, segregational freezing generates ice lenses that form at or just above the freezing front in the soil (Figure 2-1). Ice lenses actively form when the rate of heat removal approximately equals the rate of heat supply at the freezing front, and several ice lenses may form over time as the freezing front progresses deeper into the soil. The vertical displacement of the ground surface then depends on the cumulative thicknesses of the individual ice lenses.

The processes of frost heave and thaw weakening involve complex relationships between heat flow, water transport, and heaving pressure in freezing or frozen soils, as described in the following sections.

2.2 HEAT FLOW

When the air temperature is lower than the ground temperature, heat flows from the warmer soil toward the colder air. As an increasing amount of heat is removed from the soil, the soil pore water begins to freeze, and the freezing front begins to progress deeper into the soil. Because air temperatures can be much lower than ground temperatures, a large temperature gradient can occur at the air–soil interface. The rate of heat flow across this interface is directly proportional to the magnitude of the temperature gradient (Penner 1960, 1972); therefore, large temperature gradients can cause the freezing front to move rapidly through the soil nearest the surface, which leads to in situ freezing of the pore water between the soil particles and a limited amount of frost heave. However, as soon as the progression of the freezing front slows and a steady-state system is approximated, segregational freezing can become active, and ice lensing can begin, as long as the overburden pressure does not exceed the frost-heave pressure (Fredlund et al. 1991). Steady-state conditions occur when the heat removed from the soil in the vicinity of the freezing front is equal to the heat supplied from the underlying soil.

Heat flow from the underlying soil toward the freezing front occurs through both conduction and convection. While the thermal conductivity of the soil influences the rate of heat conduction, the hydraulic conductivity of the soil influences the rate of heat convection, which occurs as water moves from the underlying soil toward the freezing zone. As the incoming water arrives at the

Figure 2-1. Ice lensing in a silty soil specimen that was subjected to top-down freezing while situated in a shallow water bath.
Source: Lay (2005).

freezing front and subsequently changes from the liquid state to the solid state, its latent heat of fusion is also released, which is another important source of heat.

As long as approximately steady-state conditions continue, the thickness of an ice lens can continue to grow almost indefinitely. However, because the progression of the freezing front into the soil during the freezing process is governed by the relationship between the rate of heat removal and the rate of heat supply, dynamically changing air or soil conditions can lead to intermittent ice lensing (Fredlund et al. 1991, Penner 1962). For example, the onset of lower air temperatures or partial depletion of the water supply immediately beneath the freezing front can cause an imbalance in heat flow and, consequently, termination of ice lensing as the freezing front begins moving again deeper into the soil, continuing until the conditions needed to form an ice lens are again met. This process repeated through time results in a succession of ice lenses separated by frozen soil (Harris 1995, Penner 1962), with the ice lenses typically being thicker and more widely separated with increasing soil depth.

When the air temperature is again higher than the ground temperature, heat flows from the warmer air to the colder soil. As an increasing amount of heat is supplied to the soil, the ice begins to melt, and a thawing front begins to progress deeper into the soil. Because heat continues to flow into the frozen soil from the warmer underlying soil, a thawing front also develops along the base of the frozen soil and progresses toward the soil surface (Pylkkänen et al. 2012). Thus, during thawing, the rate of heat flow depends on the temperature gradients at both the top and the bottom of the frozen soil.

2.3 WATER TRANSPORT

Water in a freezing soil system can be supplied from several sources, including precipitation, underground aquifers, and constructed features. The proximity of the water source to the freezing front often affects the rate of frost heaving: given that heaving can take place when at least some of the individual pore spaces in the soil become filled with ice or supercooled liquid water (Hermansson and Guthrie 2005, Miller 1972), greater soil saturation (corresponding to closer water sources) yields higher rates of frost heaving (McGaw 1972). While an open system with a source of external water is required for sustained growth of ice lenses, soil does not need to be uniformly saturated for frost heaving to occur (Dirksen and Miller 1966); water in a closed system can be redistributed upward toward the freezing front, leading to frost heave even in the absence of external water (Guthrie and Hermansson 2003). Given rapid enough heat removal, the ability of water to move through the soil ultimately controls the growth of ice lenses in freezing soil.

During the freezing process, as ice begins to form in the pores between soil particles, the effective radius of the capillaries forming the unfrozen water network in the freezing soil is reduced, and the soil becomes increasingly devoid of liquid water. Both of these effects lead to the development of cryosuction, or freezing-induced negative pore water pressure. Cryosuction gradients can cause both liquid water and water vapor in the underlying soil to migrate toward the freezing front (Beskow 1947, Black and Hardenberg 1991, Guthrie et al. 2006). In this context, the two main components of cryosuction are osmotic suction and matric suction (Fredlund et al. 1991).

Osmotic suction is the force potential resulting from the presence of solutes, or ions, in the pore water (Hanks and Ashcroft 1980), where osmotic suction increases with increasing ion concentration. In the process of ice formation, ice typically nucleates near the center of the pore space and gradually expands outward. Because ice naturally forms in a pure state, ions originally present in the pore water that changes to ice are expelled into the surrounding unfrozen water. As the formation of ice continues, the expelled ions are concentrated in a decreasing amount of unfrozen water surrounding the ice within the pore space, and the ion concentration of the pore water becomes much greater than before freezing (Cary and Mayland 1972, Guthrie and Zhan 2002, Panday and Corapcioglu 1991). Because of the increased ion concentration, the freezing point of the unfrozen pore water is depressed, resulting in the presence of some supercooled liquid pore water in most frozen soil. Especially in the frozen fringe, which is the soil between the freezing front and the base of the nearest ice lens (Konrad and Morgenstern 1982), interconnected films of unfrozen water provide pathways for water movement (Figure 2-2), which can support frost heaving. In particular, differences in ion concentrations in the pore water near the freezing zone generate osmotic suction gradients, which in turn cause water to migrate from zones of lower ion concentration to zones of higher ion concentration through these unfrozen water films (Guthrie et al. 2002).

Matric suction is the force potential resulting from the presence of air–water menisci in soil and is a function of the surface tension in the contractile skin comprising the air–water meniscus. Matric suction is inherently linked to osmotic suction because of the increase in surface tension that greater ion concentrations provide (Weast et al. 1989). The magnitude of matric suction is also affected by the contact angle at which the water wets the surface of the soil particles and the radius of the air–water meniscus, which is largely determined by the sizes of the soil

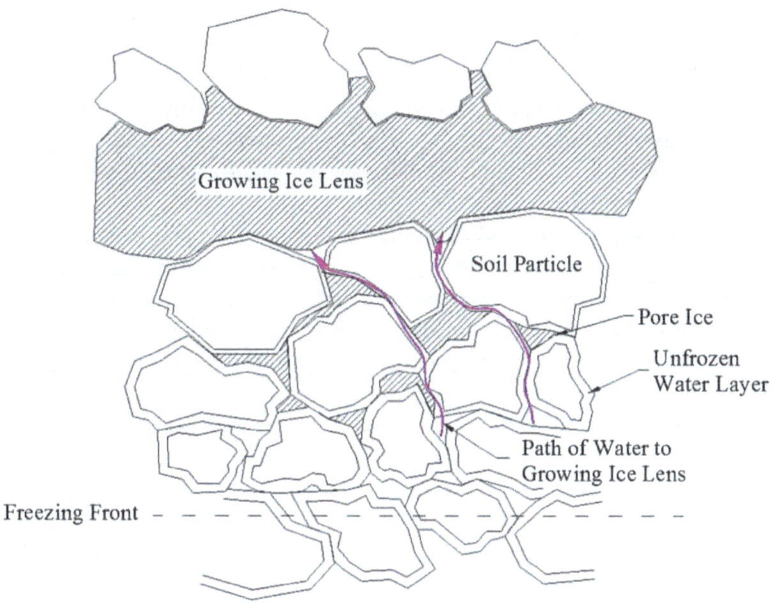

Figure 2-2. Schematic of a growing ice lens.
Source: Lay (2005).

particles. Higher surface tension, lower contact angles, and smaller radii give rise to greater matric suction in the soil (Guthrie et al. 2002). Because coarse-grained materials like gravels contain large particles, the radii of air–water menisci in these materials are also comparatively large, and the potential magnitude of matric suction is therefore low; these materials are not usually frost susceptible (Freitag and McFadden 1997). While clayey soils consist of very small particles with a potential for very high matric suction, as evidenced by considerable capillary rise potential, they are not sufficiently permeable to water to sustain a growing ice lens and are also not usually frost susceptible (Freitag and McFadden 1997). Silts, however, because they are permeable enough to allow water transport and have the required particle-size distribution to develop strong matric suction, are considered highly susceptible to frost heave (Freitag and McFadden 1997, Lay 2005).

Consistent with the laws of equilibrium, water in both liquid and vapor forms moves from areas of low total suction to areas of high total suction (Jabro 2008), where total suction is the sum of osmotic and matric suction. The rate at which water flows through a soil is affected by the magnitude of the suction gradient and the hydraulic conductivity or vapor diffusivity of the soil, which are in turn affected by the structure and degree of saturation of the soil matrix. Regarding the structure of the soil matrix, higher porosity and lower tortuosity promote higher rates of both liquid and vapor flow (Hanks and Ashcroft 1980); however, higher saturation increases liquid water flow but decreases water vapor flow.

With decreasing temperatures in a freezing soil, increasing amounts of pore water change to ice, reducing the unfrozen water content between soil particles. Although the reduced water content can cause a significant increase in matric suction, the hydraulic conductivity of the soil rapidly diminishes as freezing progresses (Hanks and Ashcroft 1980). Therefore, as liquid water migrates from warmer to colder regions within the soil matrix, and especially as it approaches an active ice lens as depicted in Figure 2-2, its path becomes increasingly tortuous and narrow until water flow is effectively terminated, usually at the base of the most recently formed ice lens.

During the thawing process, ice in the pores between soil particles begins to melt, increasing the liquid water content. Supersaturation may then result, especially in frost-susceptible soils, such as silts. Although meltwater produced by thawing of ice along the bottom of the frozen soil may drain unimpeded into the underlying soil at a rate that depends mainly on the hydraulic conductivity of the soil, meltwater produced by thawing of ice along the top of the frozen soil can become trapped above the still-frozen and nearly impermeable underlying soil. The subsequent development of positive pore water pressures under loading of the supersaturated soil can lead to the loss of interparticle friction and therefore reduce soil stiffness and strength (Doré and Zubeck 2009, Guthrie and Zhan 2002, Janoo and Berg 1990). For this reason, thaw weakening is often considered the most important frost-related damage mechanism (Berg 1988, Janoo and Berg 1990, White and Coree 1990).

2.4 HEAVING PRESSURE

Heaving of a frost-susceptible soil can continue as long as freezing temperatures persist, water is available, and the heaving pressure exceeds the overburden pressure (Hermansson and Guthrie 2005, Penner 1962). The overburden pressure at the depth of potential frost heaving depends on the self-weight of the overlying frozen soil and any surface loads, such as structures, snow, or traffic (Aitken 1963). As the pore ice grows to form an ice lens, particle-to-particle contact within the soil is disrupted. The weight of the overburden is subsequently transmitted from the soil matrix to the ice-water structure, leading to pressurization of the unfrozen water network within the soil (Cary 1987). The water necessary for ice lensing to occur will not flow into the frozen zone unless the cryosuction is greater than the counteracting overburden pressure (Cary 1987). Thus, higher overburden pressures generally correspond to less frost heave (Aitken 1963). Because the maximum heaving pressure for a given water content and temperature is mainly a function of soil type, a certain depth exists in each soil below which frost heave cannot occur; ice lenses can form only at shallower locations. Heaving pressures become negligible during thawing of frozen soil.

2.5 SUMMARY

Frost heave and thaw weakening can catastrophically effect infrastructure built on frost susceptible soils in cold regions. The degree of both frost heave and thaw weakening depends on the amount and distribution of ice that forms in the soil during freezing, which, in turn, is a function of complex relationships between heat flow, water transport, and heaving pressure in freezing or frozen soils. Understanding the mechanisms of these two phenomenon allow engineers to design solutions to mitigate damage of construction in cold regions.

References

Aitken, G. W. 1963. "Reduction of frost-heave by surcharge loading." In *Proc., Int. Conf. on Permafrost*, 319–324. West Lafayette, Indiana.

Berg, R. L. 1988. "Pavement design for frost conditions: Current and future methods." In *Proc., 4th Annual Airport Engineering/Management Conf.* Des Plains, Illinois.

Beskow, G. 1947. *Soil freezing and frost heaving with special application to roads and railroads: With a special supplement for the English translation of progress from 1935 to 1946*. Evanston, IL: Technological Institute of Northwestern Univ.

Black, P. B., and M. J. Hardenberg. 1991. *Historical perspectives in frost heave research: The early works of S. Taber and G. Beskow*. Special Rep. No. 91-23. Hanover, NH: US Army Cold Regions Research and Engineering Laboratory.

Cary, J. W. 1987. "A new method for calculating frost heave including solute effects." *Water Resour. Res.* **23** (8): 1620–1624.

Cary, J. W., and H. F. Mayland. 1972. "Salt and water movement in unsaturated frozen soil." *Soil Sci. Soc. Am.* **36** (4): 549–555.

Dirksen, C., and R. D. Miller. 1966. "Closed-system freezing of unsaturated soil." *Soil Sci. Soc. Am. J.* **30** (2): 168–173.

Doré, G., and H. K. Zubeck. 2009. *Cold regions pavement engineering.* Reston, VA: ASCE.

Fowler, A. C. 1989. "Secondary frost heave in freezing soils." *SIAM J. Appl. Math.* **49** (4): 991–1008.

Fredlund, D. G., J. K. Gan, and H. Rahardjo. 1991. "Measuring negative pore water pressures in a freezing environment." *Transp. Res. Rec.* **1307**: 291–299.

Freitag, D. R., and T. McFadden. 1997. *Introduction to cold regions engineering.* New York: ASCE.

Guthrie, W. S., and Å. Hermansson. 2003. "Frost heave and water uptake relations in variably saturated aggregate base materials." *Transp. Res. Rec.* **1821** (1): 13–19.

Guthrie, W. S., Å. Hermansson, and T. Scullion. 2002. "Determining aggregate frost susceptibility with the tube suction test." In *Cold regions engineering: Cold regions impacts on transportation and infrastructure*, 663–674. Reston, VA: ASCE.

Guthrie, W. S., Å. Hermansson, and K. H. Woffinden. 2006. "Saturation of granular base material due to water vapor flow during freezing: Laboratory experimentation and numerical modeling." In *Proc., 13th Int. Conf. on Cold Regions Engineering*, Orono, ME.

Guthrie, W. S., and H. Zhan. 2002. "Solute effects on long-duration frost heave behavior of limestone aggregates." *Transp. Res. Rec.* **1786** (1): 112–119.

Hanks, R. J., and G. L. Ashcroft. 1980. *Applied soil physics.* New York: Springer.

Harris, J. S. 1955. *Ground freezing in practice.* London: Thomas Telford.

Hermansson, Å., and W. S. Guthrie. 2005. "Frost heave and water uptake rates in silty soil subject to variable water table height during freezing." *Cold Reg. Sci. Technol.* **43** (3): 128–139.

Jabro, J. D. 2008. "Water vapor diffusion through soil as affected by temperature and aggregate size." *Transp. Porous Media* **77** (3): 417–428.

Janoo, V. C., and R. L. Berg. 1990. "Thaw weakening of pavement structures in seasonal frost areas." *Transp. Res. Rec.* **1286**: 217–233.

Kaplar, C. W. 1970. "Phenomenon and mechanism of frost heaving." *Highway Res. Rec.* **304**: 1–13.

Konrad, J. M. 1984. "Soil freezing characteristics versus heat extraction rate." In *Proc., Annual Winter Meeting*, New Orleans.

Konrad, J. M., and N. R. Morgenstern. 1982. "Prediction of frost heave in the laboratory during transient freezing." *Can. Geotech. J.* **19** (3): 250–259.

Lay, R. D. 2005. "Development of a frost heave test apparatus." M.S. thesis, Dept. of Civil and Environmental Engineering, Brigham Young Univ.

McCarthy, D. F. 1998. *Essentials of soil mechanics and foundations.* Columbus, OH: Prentice Hall.

McGaw, R. 1972. "Frost heaving versus depth to water table." *Highway Res. Rec.* **393**: 45–55.

Miller, R. D. 1972. "Freezing and heaving of saturated and unsaturated soils." *Highway Res. Rec.* **393**: 1–11.

Mitchell, J. K. 1993. *Fundamentals of soil behavior*, 2nd ed. New York: Wiley.

Panday, S., and M. Y. Corapcioglu. 1991. "Solute rejection in freezing soils." *Water Resour. Res.* **27** (1): 99–108.

Penner, E. 1960. "The importance of freezing rate in frost action in soils." *Proc. Am. Soc. Test. Mater.* **60**: 1151–1165.

Penner, E. 1962. "Ground freezing and frost heaving." *Canadian Building Digest, CBD 26, Division of Building Research, National Research Council of Canada, Ottawa.* Accessed December 12, 2019. http://web.mit.edu/parmstr/Public/NRCan/CanBldgDigests/cbd026_e.html.

Penner, E. 1972. "Influence of freezing rate on frost heaving." *Highway Res. Rec.* **393**: 56–64.

Perfect, E., R. D. Miller, and B. Burton. 1988. "Frost upheaval of overwintering plants: A quantitative field study of the displacement process." *Arct. Alpine Res.* **20** (1): 70–75.

Pylkkänen, K., H. Luomala, W. S. Guthrie, and A. Nurmikolu. 2012. "Real-time in-situ monitoring of frost depth, seasonal frost heave, and moisture in railway track structures." In *Proc., 15th Int. Conf. on Cold Regions Engineering*, Quebec.

Taber, S. 1931. "Freezing and thawing of soils as factors in the destruction of road pavements." *Public Roads* **11**: 113–132.

Weast, R. C., D. R. Lide, M. J. Astle, and W. H. Beyer, eds. 1989. *CRC handbook of chemistry and physics.* Boca Raton, FL: CRC Press.

White, T. D., and B. J. Coree. 1990. "Threshold pavement thickness to survive spring thaw." In *Proc., 3rd Int. Conf. on Bearing Capacity of Roads and Airfields*, 41–51. Trondheim, Norway.

CHAPTER 3

Effect of Frost Heave and Thaw Weakening on Roads and Other Structures

Billy Connor

Frost action in roadways and airports can generally be described as heaving, jacking, or thaw weakening. Heaving can be further broken down as uniform or differential heaving. As shown in the CRREL *Frost Action in Soils* video [Kaplar 1968, CRREL 1992, Guthrie et al. 2016, Minnesota Local Road Research Board (MN LRRB) 2013a, b] expansion of soil volume by 70% during the heaving process is not uncommon. Even so, uniform heaving typically does not cause a problem in roadways and airports. But differential heaving results in surface roughness and damage to underground utilities and culverts.

Frost jacking pushes fence posts, light posts, and even improperly designed bridge pilings out of the ground. Bridges have had to be replaced after being damaged by frost jacking. In some cases, even rocks have been expelled from the roadway surface.

Thaw weakening occurs when the ice lenses created during the frost heaving process melt. The soil layers beneath the thawed material do not allow water to readily drain, and the wet soil becomes weak. Roadways may become impassable or pavement severely damaged as the result of thaw weakening. Engineers have tried imposing load restrictions to limit the damage, but this solution is rarely effective.

Understanding the processes of heaving, jacking, and thaw weakening can help minimize or eliminate the impact of frost action on roadways. Failure to take action increases costs and adversely affects the roadway user.

3.1 FROST HEAVE

Péwé and Paige (1963) noted that frost can increase the volume of soil by 70%. This expansion is entirely caused by the migration of water to the freezing front through capillary action. If one carefully looks at the structure of heave soil, ice lensing will be evident.

In the Midwest, where frost depths in the roadway embankment are around 36 in., this implies around 25 in. of potential heave. In Alaska, where the frost can reach 15 ft in the roadway embankment, there could be 126 in. of heave. While this degree of heave is unlikely, the impacts of frost heave can be severe.

Uniform frost heave may not even be noticed. The Richardson Highway between Fairbanks and North Pole heaves more than a foot. Except near two culverts, no one even notices the heave because it is uniform.

However, differential heaving caused by non-uniform soils, a varying water supply, or varying thermal regimes can result in roughness, which can be quite severe (Figure 3-1). Even if the embankment is constructed of uniform material, a variable subgrade material may cause differential heave.

Roughness is not the only manifestation of differential heaving. In severe cases, longitudinal and transverse cracking will occur (Figure 3-2), requiring costly maintenance. Longitudinal cracking occurs when the center of the roadway heaves more than the shoulders, or vice versa. This often occurs when the roadway is widened using different material than in the original roadway.

Utility cuts may also cause differential heaving. In most cases, they are backfilled with non-frost-susceptible fill. As a result, the embankment heaves more than the cut, leaving a dip. These dips can be severe enough to damage vehicles or tires.

Culverts may have exactly the opposite effect during the freezing months. The culvert allows the frost-susceptible subgrade to freeze earlier directly below the culvert, resulting in a high spot over the culvert in early winter. As the winter progresses, the high spot disappears as the heave in the surrounding areas catches up. However, as spring thaw begins, a dip over the culvert appears, because the

Figure 3-1. Roadway damage from spring thaw weakening.
Source: Alaska Department of Transportation and Public Facilities.

Figure 3-2. Surface moisture generated by a thawing base course as a result of frost heaving.
Source: Alaska Department of Transportation and Public Facilities.

culvert area thaws faster than the adjoining roadway. By late spring or early summer, the roadway is smooth again.

3.2 FROST JACKING

As the freezing soil expands, objects in the soil will be pushed up by the freezing ground (unless they can resist the upward pressure). This can happen to fence posts, utility poles, and even bridge pilings (Domaschuk 1982). Figure 3-3 shows a fence line that has been completely jacked out of the ground. Even posts with a concrete foundation are subjected to frost jacking. The Alaska Department of Transportation and Public Facilities no longer uses this practice, because the concrete provides a large and rough surface area with good friction for the freezing soil to push the fencepost out of the ground. Driving the post deeper will reduce, but not totally eliminate, the potential for frost jacking.

Figure 3-3. Fence line affected by frost jacking despite concrete foundations.
Source: Robert Harper, Alaska University Transportation Center, University of Alaska Fairbanks.

In the 1940s, the US Army developed the Clearwater Lake area near Delta, Alaska. A wooden bridge was constructed to provide access to the recreational area. Over the years, the pilings on each end of the bridge were frost jacked. The pilings on the north end of the bridge were jacked 11 ft out of the lake bottom (Figure 3-4). Unfortunately, this was to be only one of several bridges to be damaged by frost jacking. Péwé and Paige (1963) reported that a bridge in the Goldstream Valley, near Fairbanks, was jacked 14 in. in a single season. Liverovsky and Morozov (1941) reported heave of 24 in. on a Russian bridge in a single season. The maximum frost jacking reported in Russia during a single winter was 6.5 ft. Hemstock (1953) reported frost jacking of pilings in Norman Wells, Northwest Territories, of 8 in.

The Alaska Department of Transportation and Public Facilities has repaired or replaced several bridges following frost jacking. One of the most notable is Scottie Creek Bridge, near the Canadian border, which was replaced. Riley Creek Bridge, on the Parks Highway at the entrance to Denali National Park, required remediation after the center pier jacked in the early 1970s. The pier had heaved over a foot before corrective action was taken.

Rocks have also been expelled from roadways. The phenomenon was recorded in early English literature as "breeding stones" (Manz 2011). Farmers cleared their fields of stones, only to find new ones on the surface the following spring. It was commonly believed that the pebbles themselves were the offspring of puddingstone, which was referred to as a breeding stone. The same phenomenon was reported by early farmers throughout the northern tier states: farmers cleared rocks from their fields year after year, only to find new ones every spring.

Large rocks also appeared in early wagon trails, requiring removal annually. As the roads were upgraded and paved, rocks were jacked through the surface. While this rarely occurs in modern roads, it is not entirely unheard of.

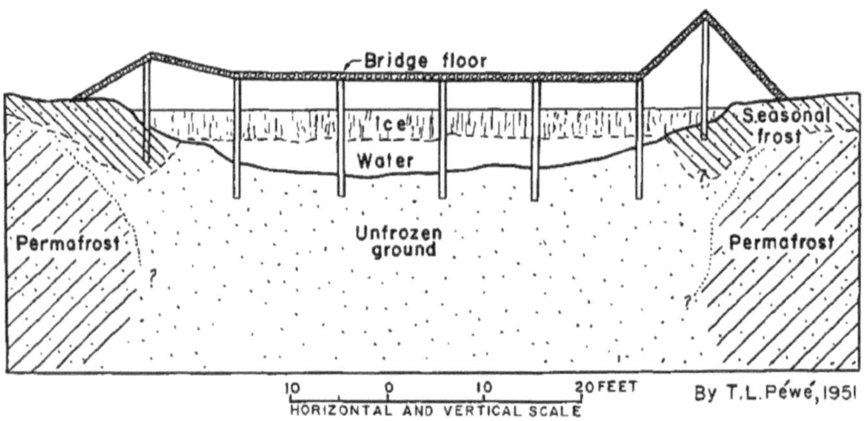

Figure 3-4. Frost jacking of bridge at Clearwater Lake near Big Delta, Alaska.
Source: Péwé and Paige (1963).

3.3 THAW WEAKENING

Thaw weakening is perhaps the most expensive manifestation of frost action. While there are no definitive estimates of the cost, it is probably in the millions of dollars annually. The cost is not only to the agency responsible for maintaining the roadway. The cost to the travelling public comes in the form of vehicle damage, longer travel times, a higher cost of commerce, and higher taxes to fund local, county, and state governments.

During the spring months, areas that have previously heaved will begin to thaw. The resulting high moisture content causes severe weakening of the roadway structure and deterioration of the surface. In many cases, the road base becomes supersaturated. While the soil moisture content cannot exceed saturation, "supersaturated" refers to the soil having higher moisture than it could have in its compacted state. The heaving action reduces the compaction of the soil, allowing more pore space for water, resulting in a very weak soil. Stubstad and Connor (1982) noted that a base course with a resilient modulus of 25 ksi (172 MPa) in summer can have a modulus as low as 1 to 3 ksi (7 to 21 MPa) in the spring.

Thaw weakening is further compounded by the frozen soil beneath the thawed soil inhibiting drainage, thereby effectively delaying the thawed soil from regaining its strength for weeks. Stubstad and Connor (1982) also showed that the thawing of the first foot of the embankment is generally the most critical period, because the shape of the deflection basin is quite pronounced, inducing high tensile strain, and therefore failure, in the asphalt layer.

Most agencies reduce the impact of thaw weakening by applying load restrictions during the spring months. There are numerous methods of determining load restrictions, including based on the judgment of Department of Transportation personnel, codified load-limit periods, or thaw depth and roadway strength measurements using a falling weight deflectometer. All have the same goal, to reduce heavy-vehicle loading during this vulnerable period and thus reduce the damage to the roadway. In some cases, such restrictions effectively shut down the movement of heavy trucks. For this reason, most states require that interstate and primary roadways be designed such that no load restrictions are required. Consequently, low-volume roadways are most subject to thaw weakening and load restrictions.

3.4 FROST ACTION ON SHALLOW FOUNDATIONS

Frost action can destroy buildings with shallow foundations through two mechanisms, frost heaving and frost jacking. Shallow foundations typically consist of spread footings, pad and post, and slab on grade. Penner and Crawford's *Frost Action and Foundations* (1983) provided case studies on the impact of frost heaving on building foundations in Canada. In one case, a cottage on the Rideau River near Ottawa was founded on concrete block, and frost heaving had distorted the building to the point that doors and windows would not function. The owner replaced the perimeter block with steel pilings. The next spring the cottage was in worse shape than it had ever been. Further investigation found that the cinder blocks supporting the center beam had not been replaced, and the center of the cottage had been lifted more than 6 in. The center blocks were then replaced with pilings, which eliminated the problem.

Luckily, larger structures are designed with foundations that are rarely affected by frost heave or frost jacking. And although frost heave does impact smaller structures more frequently, it is almost never documented. Consequently, it is impossible to estimate the cost of frost heave damage to minor structures.

References

CRREL. 1992. “Frost action in soils.” Video. Hanover, NH: Cold Regions Research and Engineering Laboratory.

Domaschuk, L. 1982. “Frost heave forces on embedded structural units.” In *Proc., 4th Canadian Permafrost Conf. on Engineering Applications in Permafrost Areas*, 487–496. Calgary, Alberta.

Guthrie, S., S. Shoop, and J. Ulring. 2016. “Frost damage in pavement: Causes and cures.” *J. Cold Reg. Eng.* **30** (3): 02516001.

Hemstock, R. 1953. *Permafrost at Norman wells, N.W.T.* Calgary, AB: Imperial Oil.

Kaplar, C. W. 1968. “Frost action in soils.” Video. Cold Regions Research and Engineering Laboratory, Hanover, NH. http://hdl.handle.net/11681/4973.

Liverovsky, A. V., and K. D. Morozov. 1941. *Construction on Permafrost: Moscow.* Translation by Meir Filch. St. Paul, MN: Military Construction Division.

Manz, L. 2011. “Frost heave.” *Geo News, 18–24.* Accessed October 31, 2019. https://www.dmr.nd.gov/ndgs/documents/newsletter/2011Summer/FrostHeave.pdf.

Minnesota Local Road Research Board (MN LRRB). 2013a. “Frost damage in pavement: Causes and cures.” Video. http://www.youtube.com/watch?feature=player_embedded&v=qnl7T5wbNjc.

MN LRRB. 2013b. “Frost damage in pavement: Causes and cures–Executive summary.” Video. https://www.youtube.com/watch?v=fkrrSys03qQ

Penner, E., and C. Crawford. 1983. *Frost action and foundations.* Ottawa, ON: National Research Council Canada.

Péwé, T. L., and R. A. Paige. 1963. *Frost heaving of piles with an example from Fairbanks, Alaska.* Washington, DC: US Government Printing Office.

Stubstad, R. N., and B. Connor. 1982. *Prediction of damage potential on Alaskan highways during spring thaw using the falling weight deflectometer.* Juneau, AK: Alaska DOT and Public Facilities.

CHAPTER 4

Assessment of Frost-Heave Susceptibility

W. Spencer Guthrie
Antti Nurmikolu

Numerous procedures have been developed to determine whether a soil is susceptible to frost heave. The availability of a variety of procedures allows engineers to select a test that has the degree of reliability and complexity appropriate for specific projects (Chamberlain 1981). The following sections describe approaches involving analysis of soil index properties, laboratory frost-heave measurements, field frost-heave measurements, and numerical modeling.

4.1 SOIL INDEX PROPERTIES

Because actual frost-heave tests require specialized equipment, the susceptibility of a soil to frost heave is instead commonly determined from various index properties used to identify and classify soils. The simplest approaches are those that evaluate aspects of the grain-size distribution of the soil (Konrad 1999), such as the percentages finer than specific grain sizes, with special consideration typically given to percentages finer than 0.02 mm and 0.075 mm (Edgers et al. 1974, Esch et al. 1981). According to one established criterion, "One should expect considerable ice segregation in non-uniform soils containing more than 3% of grains smaller than 0.02 mm, and in very uniform soils containing more than 10% smaller than 0.02 mm" (Casagrande 1931). While many agencies have applied this criterion, the additional effort required to determine the percentage finer than 0.02 mm compared to that required to determine the percentage finer than 0.075 mm may provide only a marginally stronger correlation with field performance (Edgers et al. 1974, Nurmikolu 2005). Therefore, some agencies have developed criteria based on the percentage finer than 0.075 mm (Esch et al. 1981, Homewood 2010). The ratio of the percentage finer than 0.075 mm to the percentage finer than 0.425 mm, called the dust ratio, has also been correlated with soil behavior under frost action (Esch et al. 1981). Criteria based on the coefficient of uniformity and the coefficient of curvature have also been developed (Vinson et al. 1986, Chamberlain 1981).

Other approaches supplement grain-size information with Atterberg limits test results so that soil classifications can be determined. A widely used method developed by the US Army Corps of Engineers classifies soils into four groups, F1 through F4, with F1 being the least susceptible to frost heave, and F4, the most (Freitag and McFadden 1997); in Table 4-1, these classifications have been mapped to the Unified Soil Classification System for convenience. Another prominent technique involves calculating the fines factor from grain-size information and the liquid limit of the material finer than 0.075 mm. The fines factor has been shown to be better correlated to frost-heave rate than the percentage finer than 0.02 or 0.075 mm (Vinson et al. 1986).

When further information about the frost susceptibility of soil is needed, more advanced characterization techniques can be performed. Measurements of

Table 4-1. Frost Groups Defined by the US Army Corps of Engineers.

Frost Group	*Soil Type*	*Percentage Finer Than 0.02 mm, by Weight (%)*	*Typical Soil Types Under Unified Soil Classification System*
F1	Gravelly soils	3 to 10	GW, GP, GW-GM, GP-GM
F2	(a) Gravelly soils	10 to 20	GM, GW-GM, GP-GM
	(b) Sands	3 to 15	SW, SP, SM, SW-SM, SP-SM
F3	(a) Gravelly soils	>20	GM, GC
	(b) Sands, except very fine silty sands	>15	SM, SC
	(c) Clays, plasticity index > 12	–	CL, CH
F4	(a) All silts	–	ML, MH
	(b) Very fine silty sands	>15	SM
	(c) Clays, plasticity index < 12	–	CL, CL-ML
	(d) Varved clays and other fine-grained, banded sediments	–	CL and ML; CL, ML, and SM; CL, CH, and ML; CL, CH, ML, and SM

Note: G = gravel, S = sand, M = silt, C = clay, W = well graded, P = poorly graded, H = high plasticity, L = low plasticity.
Source: NCHRP (1974).

absorption, specific surface area, cation exchange capacity, organic content, mineralogy, hygroscopicity, and electrical conductivity, for example, have been used to inform frost-susceptibility evaluations (Chamberlain 1981, Darrow et al. 2008, Guthrie et al. 2002, Nurmikolu 2005, Vinson et al. 1986). Related to both grain-size distribution and relative compaction, pore-size distribution has also been examined. Laboratory methods for assessing aspects of pore-size distribution include capillary rise measurement, mercury intrusion, and a variety of procedures relating water content to the development of cryosuction or freezing-induced negative pore water pressure (Chamberlain 1981).

4.2 LABORATORY FROST-HEAVE MEASUREMENTS

Several laboratory frost-heave test devices have been developed, with varying levels of technical sophistication, for performing standard tests or for simulating the conditions of a particular site (Darrow et al. 2008, Guthrie et al. 2002, Henry and Holtz 2001, Hermansson and Guthrie 2005, Jessberger and Carbee 1970, Jones 1980, Kaplar 1970, Konrad and Morgenstern 1980, Lay 2005, Loch and Kay 1978, Nurmikolu 2005, Penner 1960, Taber 1931, Yong et al. 1984). The general design of frost-heave test devices has remained the same since the first documented laboratory frost-heave testing was conducted (Taber 1931). Almost without exception, the method has been to use a cylindrical specimen, provide water and heat sources at the bottom, insulate the sides, and expose the top to freezing temperatures (Dirksen and Miller 1966, Kaplar 1970), as illustrated in Figure 4-1. However, many improvements have been made in testing protocols to facilitate more realistic approximations of field conditions and to enable efficient collection of meaningful test data.

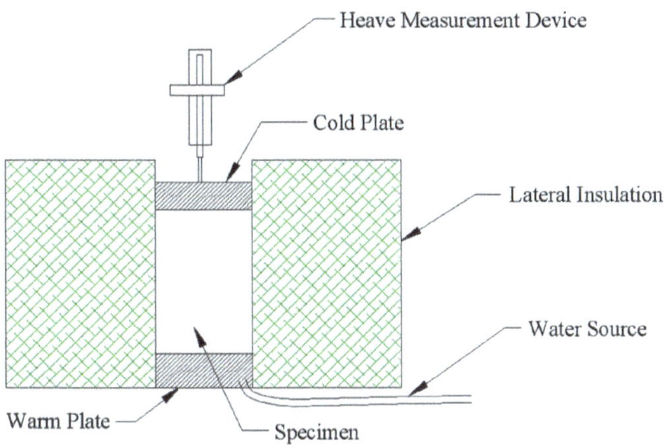

Figure 4-1. Generic frost-heave device setup.

Regarding data collection, three parameters that are nearly universally measured in laboratory frost-heave tests are surface heave, temperature distribution, and water intake (Konrad 1994). A parameter less commonly measured is heaving pressure. Although quantifying heaving pressure can be an important aspect of frost-heave testing, it is much harder to measure than the other parameters. Surface heave and temperature profiles are typically measured continuously throughout the duration of testing by using linear variable differential transformers, while water uptake can be monitored with a differential pressure transducer (Darrow et al. 2008); but the most common way to measure water uptake is to manually weigh the specimens before and after the frost-heave test or at set intervals during the testing (Guthrie et al. 2002).

An important aspect of laboratory frost-heave testing is controlling the thermal gradient across the specimen, which should ideally represent the field conditions of interest. The thermal gradient is controlled by the heat source and heat sink used in the test device. For the heat source, a warm plate can be used to regulate the temperature of the water as it enters the specimen, or the specimen may be placed in a water bath maintained at a specific temperature by using thermostat-controlled heat tape, for example. For the heat sink, a cold plate can be used to remove heat from tested specimens (Nurmikolu 2005), or the entire frost-heave test apparatus can be placed in a cold chamber, allowing the circulation of freezing air to remove heat from the specimen surfaces (Guthrie et al. 2002). When cold air is used as the heat sink, the surface temperature of the specimen is usually warmer than the air in the cold chamber. Depending on the limits of temperature control for the heat source and heat sink, different specimen heights may be required to achieve specific thermal gradients. Compared to taller specimens, shorter specimens require a smaller temperature difference between the heat source and heat sink to maintain a given thermal gradient. Laboratory frost-heave testing has used thermal gradients ranging from approximately 20 °C/m to 150 °C/m.

Moisture conditioning of specimens before and during laboratory frost-heave testing should also ideally represent the field conditions of interest. In some cases, the specimen may be fully saturated prior to testing and maintained in a saturated state throughout the testing; in other cases, the specimen may be only initially saturated or not saturated at all and then provided a free water source at the bottom during the testing period (Guthrie et al. 2002, Nurmikolu 2005). The ability to evaluate the frost-heave behavior of specimens conditioned with saline water is also sometimes beneficial, as some soils may have high ion concentrations in the pore water (Guthrie and Zhan 2002).

In nature, the lateral expansion of a freezing soil is prevented by the soil adjacent to it, and frost heave then generally occurs along the vertical axis. Therefore, the container or mold in which the specimen is tested must have adequate tensile strength to prevent lateral deformations during freezing. Furthermore, the container or mold should also minimize adfreezing, in which the freezing soil adheres to the inside surface of the container or mold as a result of ice formation at that interface. This creates an unnatural dragging force that opposes the frost heave of the specimen. One effective remediation method for this problem is the use of a multi-ring testing container (Hasson and Irwin 1988), and this is the most commonly used; but rubber membranes have also been effective (Chamberlain 1981; Johnson et al. 1986).

Although some frost-heave tests are not configured to evaluate overburden pressure, the ability to apply variable surcharges to simulate the expected field behavior of deeper layers is useful. Metal plates are commonly used to produce the desired pressure (Lay 2005).

For expediency in testing numerous specimens, a frost-heave testing device should accommodate more than one specimen at a time. Testing of replicate specimens is recommended for statistically substantiating test results; in this case, the conditions imposed on all specimens must be identical. Therefore, great care must be taken to ensure that variability in temperatures, water supply, and other such parameters within the frost-heave testing device is minimized.

If a specimen exhibits different frost-heave behavior in an initial test than in subsequent tests, application of more than one freeze–thaw cycle might be warranted. The laboratory frost-heave test protocol developed by the US Army Corps of Engineers' Cold Regions Research Engineering Laboratory requires two cycles (Edgers et al. 1974); and other test methods require three or more (Chamberlain 1981).

When evaluation of thaw weakening is desired, specimens can be subjected to thawing conditions after frost-heave testing, and then a strength or stiffness test, such as the California bearing ratio test, can be performed (Edgers et al. 1974). In some applications, characterizing the degree of thaw weakening may be considered the most important aspect of a frost-susceptibility classification (Chamberlain 1981).

4.3 FIELD FROST-HEAVE MEASUREMENTS

In situ frost-heave tests are usually designed for observing long-term trends in the field under prevailing conditions. In situ tests are particularly useful for characterizing field conditions for simulation in the laboratory; differences between natural and laboratory conditions account for many of the unsatisfactory results of laboratory experimentation (Penner 1960). Because environmental conditions cannot be controlled in the field, measuring the air temperature, soil temperature, and soil water content together with frost heave is a recommended practice.

Temperatures can be measured using a variety of sensors, including thermocouples, thermistors, and semiconductors (Pylkkänen et al. 2012). When used to measure soil temperature, the sensors should be configured to ensure proper thermal contact with the soil. A common approach is to mount the sensors at the desired vertical spacing in the wall of a plastic tube and then insert the tube into a hole made at the desired location (Guthrie et al. 2004, Pylkkänen et al. 2012).

Removable frost-depth indicator tubes incorporating a dye that turns colorless when cooled to 0 °C can also be used when only the depth of frost penetration is of interest (Edgers et al. 1974). Heat flux sensors can also be installed to measure heat flow directly (Fukuda and Nakagawa 1985).

Changes in water content are commonly measured using time-domain reflectometry and dielectric-type sensors embedded directly in the soil (Pylkkänen et al. 2012). Such subsurface sensors can be useful to track the depth of shallow water tables and to assess the response of soil layers to precipitation, for example, which may be measured separately. When absolute measurements of water content are needed, the sensors must be calibrated; and soil samples used for this purpose must be compacted to the same density as that in the field to achieve the highest accuracy. Further characterization of moisture effects may be accomplished with sensors for matric suction, for example (Fukuda and Nakagawa 1985).

Measurements of maximum frost heave in the field may be useful as an index for the amount of damage that occurred during a given winter and may also be easily related to a performance criterion; however, the rate of frost heaving for a given soil or site may be more useful for comparing data collected during freezing seasons of different lengths, because the rate of frost heaving is less sensitive to the duration of the freezing period than to the maximum frost heave (Edgers et al. 1974). Both the frost depth and the frost-heaving rate at a given site will be affected by the type of surface cover at the site, where snow and turf typically behave as insulation. Evaluations of thaw weakening in the spring can be performed using plate load tests (Edgers et al. 1974).

When field test sites can be adequately characterized, the observed frost-heave data can be extremely valuable for validating the results of laboratory frost-heave tests or the suitability of specific soil index properties for assessing frost-heave susceptibility (Chamberlain 1981). Field data can also be valuable for validating numerical models developed to simulate frost-heave behavior (Fukuda and Nakagawa 1985).

4.4 NUMERICAL MODELING

Numerical modeling of frost-heave behavior can be used either to explain or to predict frost heave (Han and Goodings 2006). Numerical models may be empirical or theoretical. Empirical models usually include various combinations of soil index properties that provide statistically significant correlations with selected results of frost-heave testing, including but not limited to heave ratio (ratio of frost heave to frost penetration) and total heave after a particular freezing duration (Chamberlain 1981).

Theoretical models are usually based on fundamental equations representing physical mechanisms involved in the frost-heave process. Systems of partial differential equations are solved simultaneously in computer programs using finite difference or finite element approximations (Guymon et al. 1980). One-dimensional modeling is the most common, although two- or three-dimensional approaches may be more appropriate in many situations (Guymon et al. 1981). The most advanced models will allow estimation of the coupled flow of heat, water, and solutes in freezing, unsaturated soil with variable initial conditions and time-dependent boundary conditions, including external loading (Cary 1987). Heat transfer in the form of conduction and convection and water movement in the form of liquid and vapor may be simulated (Guymon et al. 1981). Convection and diffusion of solutes and the effects of solute concentration on freezing-point depression, unfrozen water film thickness, and hydraulic conductivity may also be simulated (Henry 1988). Simpler models frequently neglect water vapor movement and the potential effects of solutes on frost-heave susceptibility because of the greater mathematical complexity required to address these aspects and because many field soils have low solute concentrations within the freezing zone (Henry 1988). Nonetheless, solutes can play an important role in frost-heave behavior (Guthrie and Zhan 2002).

A complete numerical model may thus be able to compute temperature, liquid water, ice, and solute concentration with depth over time and consequently also predict frost-heaving rate, total frost heave, heaving pressure, and locations of ice lenses in a simulated soil. In some cases, calculations of internal stresses will also be performed (Mu and Ladanyi 1987). Predictions of thaw settlement may also be generated by numerical models (Yang and Goodings 1998).

While advanced numerical models offer many possible options for analysis of frost heave, determining the numerous input values for the models requires significant effort in most cases. Indeed, obtaining the unsaturated flow parameters needed for many numerical models requires equipment seldom available in a standard geotechnical laboratory; consequently, when possible, default or backcalculated values are commonly used (Yang and Goodings 1998). For example, if frost heave is known for a given site for one set of conditions, selected input values can be backcalculated using an appropriate numerical model and then used in subsequent predictions of frost heave under different boundary conditions (Yang and Goodings 1998). Determining the types and magnitudes of adjustments that should be made in the backcalculation process usually requires informed judgment or experience with the given numerical model. In particular, modelers should recognize that efforts to simultaneously calibrate a model against two or more parameters are usually unsuccessful in one or more respects. Ultimately, the accuracy of a numerical model depends as much on correct simulation of the mechanics as it does on the selection of appropriate input values.

Deterministic solutions to numerical models, empirical or theoretical, may be presented in terms of confidence bands based on given or anticipated variations in parameters and boundary conditions (Guymon et al. 1981). Quantifying the sensitivity of models to variation in specific inputs can also be helpful in interpreting the output (Yang and Goodings 1998).

4.5 SUMMARY

The many available methods to determine frost susceptibility of soil allow the practitioner to choose the appropriate test based on project needs. Because actual frost-heave tests require specialized equipment, the susceptibility of a soil to frost heave is instead commonly determined from various soil index properties used to identify and classify soils. The simplest approaches are those that evaluate aspects of the grain-size distribution of the soil; when further information about the frost susceptibility of soil is needed, more advanced characterization techniques can be used. Several laboratory frost-heave test devices have been developed, with varying levels of technical sophistication, for performing standard tests or for simulating the conditions of a particular site. Almost without exception, the method has been to use a cylindrical specimen, provide water and heat sources at the bottom, insulate the sides, and expose the top to freezing temperatures.

Regarding data collection, three parameters that are nearly universally measured in laboratory frost-heave tests are surface heave, temperature distribution, and water intake; less commonly measured is heaving pressure. In situ frost-heave tests are usually designed for observing long-term trends in the field under prevailing conditions. Because environmental conditions cannot be controlled in the field, measuring the air temperature, soil temperature, and soil water content together with frost heave is a recommended practice. Numerical modeling of frost-heave behavior can be used to either explain or predict frost heave. Numerical models may be empirical or theoretical.

References

Cary, J. W. 1987. "A new method for calculating frost heave including solute effects." *Water Resour. Res.* **23** (8): 1620–1624.

Casagrande, A. 1931. "Discussion of frost heaving." In Vol. **11** of *Proc., 11th Annual Meeting of the Highway Research Board*, Washington, DC, 168–172.

Chamberlain, E. J. 1981. "Comparative evaluation of frost-susceptibility tests." *Transp. Res. Rec.* **809**: 42–52.

Darrow, M. M., S. L. Huang, Y. Shur, and S. Akagawa. 2008. "Improvements in frost heave laboratory testing of fine-grained soils." *J. Cold Reg. Eng.* **22** (3): 65–78.

Dirksen, C., and R. D. Miller. 1966. "Closed-system freezing of unsaturated soil." *Soil Sci. Soc. Am. J.* **30** (2): 168–173.

Edgers, L., L. Bedingfield, and N. Bono. 1974. "Field evaluation of criteria for frost susceptibility of soils." *Transp. Res. Rec.* **1190**: 73–85.

Esch, D. C., R. L. McHattie, and B. Connor. 1981. "Frost-susceptibility ratings and pavement structure performance." *Transp. Res. Rec.* **809**: 27–34.

Freitag, D. R., and T. McFadden. 1997. *Introduction to cold regions engineering.* New York: ASCE.

Fukuda, M., and S. Nakagawa. 1985. "Numerical analysis of frost heaving based upon the coupled heat and water flow model." In *Proc., 4th Int. Symp. on Ground Freezing*, Sapporo, Japan.

Guthrie, W. S., B. J. Blankenagel, and C. Nielsen. 2004. "Low-intensity infrared heating for frozen ground preparation in civil construction." In *Proc., 12th Int. Specialty Conf. on Cold Regions Engineering*, Edmonton, Canada.

Guthrie, W. S., Å. Hermansson, and T. Scullion. 2002. "Determining aggregate frost susceptibility with the tube suction test." In *Cold regions engineering: Cold regions impacts on transportation and infrastructure*, 663–674. Reston, VA: ASCE.

Guthrie, W. S., and H. Zhan. 2002. "Solute effects on long-duration frost heave behavior of limestone aggregates." *Transp. Res. Rec.* **1786** (1): 112–119.

Guymon, G. L., R. L. Berg, T. C. Johnson, and T. V. Hromadka, II. 1981. "Results from a mathematical model of frost heave." *Transp. Res. Rec.* **809**: 2–6.

Guymon, G. L., T. V. Hromadka, II, and R. L. Berg. 1980. "A one-dimensional frost heave model based upon simulation of simultaneous heat and water flux." *Cold Reg. Sci. Technol.* **3** (2–3): 253–262.

Han, S. J., and D. J. Goodings. 2006. "Practical model of frost heave in clay." *J. Geotech. Geoenviron. Eng.* **132** (1): 92–101.

Hasson, P. F. X., and L. H. Irwin. 1988. "Comparison of freezing cells used in the determination of soil frost susceptibility." *ASTM Geotech. Test. J.* **11** (3): 218–220.

Henry, K. 1988. *Chemical aspects of soil freezing.* CRREL Rep. No. 88-17. Hanover, NH: USACE, Cold Regions Research Engineering Laboratory.

Henry, K. S., and R. D. Holtz. 2001. "Geocomposite capillary barriers to reduce frost heave in soils." *Can. Geotech. J.* **38** (4): 678–694.

Hermansson, Å., and W. S. Guthrie. 2005. "Frost heave and water uptake rates in silty soil subject to variable water table height during freezing." *Cold Reg. Sci. Technol.* **43** (3): 128–139.

Homewood, A. R. 2010. "Effects of thermal gradient and fines content on frost heave of an Alaska base material." M.S. thesis, Dept. of Civil and Environmental Engineering, Brigham Young Univ.

Jessberger, H. L., and D. L. Carbee. 1970. "Influence of frost action on the bearing capacity of soils." *Highway Res. Rec.* **304**: 14–26.

Johnson, T. C., R. L. Berg, and A. DiMillio. 1986. "Frost action predictive techniques: An overview of research results." *Transp. Res. Rec.* **1089**: 147–161.

Jones, R. H. 1980. "Frost heave of roads." *Q. J. Eng. Geol.* **13** (2): 77–86.

Kaplar, C. W. 1970. "Phenomenon and mechanism of frost heaving." *Highway Res. Rec.* **304**: 1–13.

Konrad, J. M. 1994. "Sixteenth Canadian geotechnical colloquium: Frost heave in soils: Concepts and engineering." *Can. Geotech. J.* **31** (2): 223–245.

Konrad, J. M. 1999. "Frost susceptibility related to soil index properties." *Can. Geotech. J.* **36** (3): 403–417.

Konrad, J. M., and N. R. Morgenstern. 1980. "A mechanistic theory of ice lens formation in fine-grained soils." *Can. Geotech. J.* **17** (4): 473–486.

Lay, R. D. 2005. "Development of a frost heave test apparatus." M.S. thesis, Dept. of Civil and Environmental Engineering, Brigham Young Univ.

Loch, J. P., and B. D. Kay. 1978. "Water redistribution in partially frozen, saturated silt under several temperature gradients and overburden loads." *Soil Sci. Soc. Am. J.* **42** (3): 400–406.

Mu, S., and B. Ladanyi. 1987. "Modelling of coupled heat, moisture and stress field in freezing soil." *Cold Reg. Sci. Technol.* **14** (3): 237–246.

NCHRP (National Cooperative Highway Research Program). 1974. *Roadway design in seasonal frost areas: NCHRP synthesis 26.* Washington, DC: NCHRP, Transportation Research Board, National Research Council.

Nurmikolu, A. 2005. "Degradation and frost susceptibility of crushed rock aggregates used in structural layers of railway track." Ph.D. dissertation, Tampere Univ. of Technology.

Penner, E. 1960. "The importance of freezing rate in frost action in soils." *Proc. Am. Soc. Test. Mater.* **60**: 1151–1165.

Pylkkänen, K., H. Luomala, W. S. Guthrie, and A. Nurmikolu. 2012. "Real-time in-situ monitoring of frost depth, seasonal frost heave, and moisture in railway track structures." In *Proc., 15th Int. Conf. on Cold Regions Engineering*, Quebec.

Taber, S. 1931. "Freezing and thawing of soils as factors in the destruction of road pavements." *Public Roads* **11** (6): 113–132.

Vinson, T. S., F. Ahmad, and R. Rieke. 1986. "Factors important to the development of frost heave susceptibility criteria for coarse-grained soils." *Transp. Res. Rec.* **1089**: 124–131.

Yang, D., and D. J. Goodings. 1998. "Predicting frost heave using FROST model with centrifuge models." *J. Cold Reg. Eng.* **12** (2): 64–83.

Yong, R. N., P. Boonsinsuk, and A. E. Tucker. 1984. "A study of frost-heave mechanics of high-clay content soils." *J. Energy Resour. Technol.* **106** (4): 502–508.

CHAPTER 5

Frost Action and Climate Change

Travis Logan
Jean-Pascal Bilodeau
Karen Henry

The fifth assessment report (AR5) of the Intergovernmental Panel on Climate Change states that climate warming is unequivocal and that it is "extremely likely" that human influence has been the dominant cause of the observed warming since the mid-twentieth century (IPCC 2013). In fact, according to the most recent data analyzed by the National Oceanic and Atmospheric Administration (NOAA 2015), the 2015 year-to-date (January–August) global mean land temperature was the highest in the 1880–2015 record. AR5 also indicates that each of the last three decades has been successively warmer; and each has also been warmer than any decade since 1850 (Figure 5-1). Increases in ocean temperatures and sea level, and decreases in sea-ice extent and snow, have also been observed. Between the mid-1980s and the mid-2000s, as global temperatures change, permafrost temperatures have increased by 3 °C in parts of Northern Alaska. Permafrost studies in Russia also indicate a considerable reduction in both permafrost thickness and areal extent between 1975 and 2005. Precipitation regimes have also changed, with some regions displaying increases and others, decreases.

Climate projections indicate that these trends are likely to continue through the twenty-first century (Collins et al. 2013). By the end of the century, the mean increase in global temperature (calculated as a mean of an ensemble of global climate model simulations) is projected to reach 1.8 °C following RCP 4.5, and up to 3.7 °C following RCP 8.5. Note: Of the four "representative concentration pathways," RCP 4.5 and 8.5 in particular have been used widely to assess the potential impacts of climate change as they represent contrasting levels of greenhouse gas mitigation efforts. RCP 4.5 is a moderate scenario, with radiative forcing increasing relative to preindustrial conditions but then stabilizing in the mid-twenty-first century. RCP 8.5 is a high-emissions scenario representing little

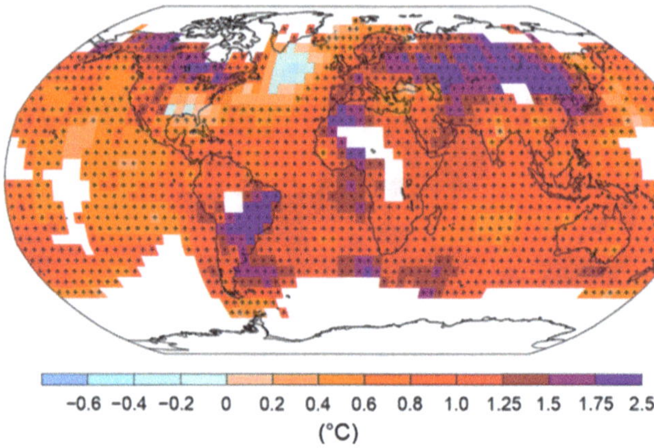

Figure 5-1. Observed change in temperature from 1900 to 2012. The + signs indicate where the trend is significant at the 10% level; white boxes indicate where there were not enough data for a robust estimate.
Source: IPCC (2013).

or no mitigation effort, with a more or less linear increase in radiative forcing until 2100 (Moss et al. 2010).

The AR5 results further indicate that changes in global temperature and precipitation regimes may have large effects on the cryosphere, with Arctic sea ice and snow cover, near-surface permafrost, and glacier ice all projected to decrease. More specifically, in a high-emissions scenario, the areal extent of Northern Hemisphere spring snow cover is projected to shrink by up to 25% (medium confidence), and that of near-surface permafrost by up to 81% (medium confidence).

Projections for North America (and other areas of the globe) are that future increases in temperature will not be uniform for all regions (Collins et al. 2013, Maloney et al. 2014). Indeed, simulated changes in temperature show a clear amplification of the temperature-change signal in the higher latitudes (Maloney et al. 2014). Changes in temperature will likely vary between seasons, with increases in winter temperatures generally being larger, and with a stronger north–south gradient, than in summer. Spatial and seasonal variability is also apparent in projections for other climate variables, such as precipitation indices, snow cover, evapotranspiration, and surface runoff (Maloney et al. 2014).

The following sections present projected changes in winter climate conditions over North America and specific indices that are important for frost action. While the list of included indicators is not exhaustive, this portrait provides a base of information for assessing the potential impacts of climate change on frost-action processes. It should therefore serve as a relatively robust starting point for more-detailed future research avenues.

5.1 NORTH AMERICAN CLIMATE CHANGE PORTRAIT OF WINTER AND FROST ACTION VARIABLES

5.1.1 Climate Indices and Determination of North American Subregions

Table 5-1 presents future climate changes in North America. The values are from a combination of observed and projected future data sets. Observed conditions of mean climate indicators deemed important in influencing frost action in frozen ground were calculated over North America for 1981 to 2010 using the Global Meteorological Forcing Dataset for land surface modeling (Sheffield et al. 2006). To produce subregions, observed mean conditions for the $0.25° \times 0.25°$ grid cells were used in an unsupervised *k*-means classification to divide the study area into relatively homogeneous subregions (Figure 5-2). For the purpose of this discussion, it is likely that projected changes in snow cover variables (not included in the climate analysis) correlate highly with several of the presented indices (e.g., winter length and total annual snowfall).

5.1.2 Future Projections of Climate Indexes

A robust portrait of projected change was produced by using a large ensemble of future simulations, such as the World Climate Research Programme's Coupled Model Intercomparison Project Phase 5 (CMIP5) multi-model data set (Taylor et al. 2012). This enabled the assessment of projected future changes in mean conditions while providing a robust evaluation of various sources of uncertainty in the projections, such as climate model uncertainty and emissions uncertainty (Charron 2014).

Projected changes for North American subregions were determined using the NASA Earth Exchange Global Daily Downscaled Projections (NEX-GDDP) data set. It is composed of downscaled high-resolution climate scenarios from 21 CMIP5 global climate model runs across RCP 4.5 and RCP 8.5. Relative future change values (Δ - delta) are presented as either the difference (expressed in climate index units) or the ratio (expressed in percentage change) between the 30-year future average and historical values. Ensemble results by RCP were summarized by calculating the mean of all delta values (the ensemble mean) for a given climate index and future horizon. Uncertainty in future changes is expressed as the standard deviation of these same delta values. (Each of the NEX-GDDP simulations includes daily maximum temperature, minimum temperature, and precipitation for 1950–2100, with a spatial resolution of 0.25°, from which the various climate indices listed in Table 5-1 can be derived. For each simulation, climate indices were calculated for a 30-year historical period [1986–2005] and two future horizons [2041–2070 and 2071–2100] to generate delta values.)

Table 5-1. Observed Climatic Normals (1981–2010) and Projected Futures Changes (Δ) of Winter and Frost-Action Climate Indicators over North American Subregions.

		Observed		2041–2070		2071–2100	
	Climate Variable	*1981–2000*	*Δ units*	*Δ RCP 4.5*	*Δ RCP8.5*	*Δ RCP 4.5*	*Δ RCP8.5*
High Arctic	*Tavg DJF* (°C)	−29.3	°C	+4.9 ± 1.6	+6.9 ± 1.9	+6.0 ± 2.0	+11.4 ± 2.8
	Prec DJF (mm)	20.7	%	+28.0 ± 11.1	+32.9 ± 24.6	+35.1 ± 13.1	+61.8 ± 32.8
	SnwFrac DJF (%)	100	%	0.0 ± 0.0	0.0 ± 0.1	0.0 ± 0.1	−0.5 ± 1.0
	SnwFall ANN (mm SWE)	98.3	%	+4.8 ± 5.4	+1.0 ± 16.8	+5.0 ± 5.1	+0.2 ± 18.7
	AFI (degree-days)	5366.6	degree-days	−1075 ± 334.7	−1479 ± 372.8	−1312 ± 407.1	−2337 ± 530.9
	ContFrz (days)	270.2	days	−17.1 ± 6.8	−24.4 ± 8.3	−21.0 ± 8.1	−41.3 ± 13.5
	FallFrStart (j.d.)	16-Aug	days	+14.8 ± 5.5	+20.6 ± 6.4	+18.0 ± 6.1	+33.3 ± 9.5
Northeastern Canada	*Tavg DJF* (°C)	−13.2	°C	+2.9 ± 0.9	+4.0 ± 0.9	+3.5 ± 1.0	+6.2 ± 1.2
	Prec DJF (mm)	216.4	%	+12.0 ± 5.0	+18.4 ± 12.3	+16.2 ± 5.9	+27.8 ± 12.2
	SnwFrac DJF (%)	96.9	%	−6.9 ± 2.3	−10.2 ± 2.9	−8.8 ± 2.9	−20.2 ± 5.4
	SnwFall ANN (mm SWE)	383.8	%	−7.0 ± 5.7	−9.5 ± 6.6	−8.5 ± 6.3	−21.5 ± 8.5
	AFI (degree-days)	1625.8	degree-days	−448 ± 133.2	−612 ± 134.5	−546 ± 148.1	−922 ± 155.5
	ContFrz (days)	163.8	days	−20.3 ± 7.7	−29.2 ± 9.3	−25.7 ± 9.2	−50.8 ± 14.1
	FallFrStart (j.d.)	10-Oct	days	+11.2 ± 3.7	+15.9 ± 4.6	+14.0 ± 4.6	+24.9 ± 6.1
Northern Alpine	*Tavg DJF* (°C)	−7.1	°C	+2.1 ± 0.7	+2.8 ± 0.8	+2.5 ± 0.7	+4.6 ± 1.1
	Prec DJF (mm)	288.6	%	+13.0 ± 6.8	+19.8 ± 15.1	+16.7 ± 7.9	+32.2 ± 17.7
	SnwFrac DJF (%)	89.9	%	−12.0 ± 4.1	−17.7 ± 6.3	−15.9 ± 4.8	−31.8 ± 9.0
	SnwFall ANN (mm SWE)	441.8	%	−14.5 ± 5.6	−19.9 ± 10.5	−18.7 ± 7.4	−36.9 ± 11.6
	AFI (degree-days)	942.1	degree-days	−355 ± 90.6	−445 ± 106.8	−415 ± 98.6	−652 ± 108.1
	ContFrz (days)	157.8	days	−32.8 ± 8.8	−44.2 ± 11.5	−40.2 ± 10.9	−73.1 ± 16.0
	FallFrStart (j.d.)	25-Sep	days	+12.0 ± 3.3	+16.4 ± 5.4	+14.8 ± 3.5	+25.3 ± 5.5

Northern Plains/ Boreal	*Tavg DJF* (°C)	−18.6	°C	+3.5 ± 1.0	+4.7 ± 1.2	+4.2 ± 1.2	+7.5 ± 1.7
	Prec DJF (mm)	64.8	%	+17.7 ± 6.5	+20.6 ± 15.3	+21.7 ± 7.0	+36.4 ± 18.7
	SnwFrac DJF (%)	99.6	%	−0.7 ± 0.7	−1.4 ± 1.1	−1.1 ± 1.0	−5.0 ± 3.5
	SnwFall ANN (mm SWE)	153.5	%	+2.9 ± 4.0	+0.8 ± 12.4	+3.0 ± 4.4	−1.8 ± 14.9
	AFI (degree-days)	2523.9	degree-days	−624 ± 163.2	−836 ± 183.2	−750 ± 196.4	−1295 ± 255.8
	ContFrz (days)	192.2	days	−17.1 ± 5.3	−23.4 ± 6.4	−21.1 ± 6.4	−41.2 ± 10.8
	FallFrStart (j.d.)	26-Sep	days	+9.2 ± 2.5	+13.1 ± 3.2	+11.5 ± 2.9	+20.9 ± 3.8
Eastern USA	*Tavg DJF* (°C)	2.6	°C	+2.1 ± 0.8	+2.7 ± 0.8	+2.4 ± 0.8	+4.2 ± 1.1
	Prec DJF (mm)	257.4	%	+6.9 ± 8.3	+9.5 ± 11.3	+9.7 ± 9.0	+15.1 ± 17.2
	SnwFrac DJF (%)	31.6	%	−8.3 ± 3.6	−10.3 ± 3.6	−9.5 ± 3.8	−14.0 ± 3.9
	SnwFall ANN (mm SWE)	89	%	−40.5 ± 15.3	−50.8 ± 14.7	−46.0 ± 15.7	−69.8 ± 14.2
	AFI (degree-days)	56.8	degree-days	−41 ± 12.2	−48 ± 9.6	−46 ± 11.5	−56 ± 9.2
	ContFrz (days)	24	days	−14.0 ± 3.8	−17.0 ± 3.1	−15.8 ± 3.8	−22.0 ± 2.6
	FallFrStart (j.d.)	15-Nov	days	+8.8 ± 3.1	+7.4 ± 7.4	+9.6 ± 4.3	+3.3 ± 18.8
Interior USA	*Tavg DJF* (°C)	−3.6	°C	+2.3 ± 0.8	+3.1 ± 1.0	+2.8 ± 1.0	+4.9 ± 1.4
	Prec DJF (mm)	75.5	%	+10.6 ± 4.7	+13.5 ± 12.7	+14.4 ± 4.7	+23.9 ± 15.0
	SnwFrac DJF (%)	69.9	%	−12.9 ± 4.7	−17.1 ± 6.4	−15.9 ± 6.1	−28.5 ± 8.3
	SnwFall ANN (mm SWE)	78.5	%	−22.2 ± 9.7	−29.7 ± 16.7	−26.7 ± 11.3	−48.3 ± 18.2
	AFI (degree-days)	447.5	degree-days	−211 ± 54.5	−260 ± 60.5	−242 ± 64.8	−358 ± 60.5
	ContFrz (days)	87.3	days	−26.1 ± 7.9	−34.2 ± 9.3	−31.3 ± 9.6	−52.3 ± 11.4
	FallFrStart (j.d.)	19-Oct	days	+8.5 ± 2.3	+11.5 ± 3.7	+10.8 ± 2.8	+18.0 ± 4.2

(Continued)

Table 5-1. Observed Climatic Normals (1981–2010) and Projected Futures Changes (Δ) of Winter and Frost-Action Climate Indicators over North American Subregions. (Continued)

		Observed		2041–2070		2071–2100	
	Climate Variable	*1981–2000*	*Δ units*	*Δ RCP 4.5*	*Δ RCP8.5*	*Δ RCP 4.5*	*Δ RCP8.5*
Pacific	*Tavg DJF* (°C)	1.4	°C	+2.0 ± 0.5	+2.5 ± 0.7	+2.3 ± 0.6	+4.1 ± 0.9
	Prec DJF (mm)	510.1	%	+9.7 ± 5.4	+20.0 ± 32.5	+12.9 ± 6.2	+29.4 ± 33.5
	SnwFrac DJF (%)	31.4	%	−9.4 ± 2.7	−10.3 ± 3.6	−10.3 ± 3.0	−13.1 ± 2.8
	SnwFall ANN (mm SWE)	165.9	%	−62.8 ± 12.1	−67.7 ± 19.8	−68.9 ± 11.9	−87.4 ± 9.9
	AFI (degree-days)	41.8	degree-days	−38 ± 8.3	−39 ± 9.9	−39 ± 9.4	−44 ± 7.4
	ContFrz (days)	28.5	days	−21.4 ± 4.3	−24.0 ± 4.3	−23.3 ± 4.2	−29.3 ± 2.1
	FallFrStart (j.d.)	02-Nov	days	+2.2 ± 6.4	+5.9 ± 9.1	+7.3 ± 10.8	+10.1 ± 12.0

Notes: Future projections indicate the ensemble mean change and model uncertainty (±1 σ) between individual simulations of the NEX-GDDP data set results ($n = 21$ for each RCP).

Tavg DJF (°C) = Mean daily maximum temperature in December, January, and February.

Prec DJF (mm) = Total amount of liquid and solid precipitation in December, January, and February.

SnwFrc DJF (%) = Percentage of *Prec DJF* falling as snow.

SnwFall ANN (mm snow water equivalent [SWE]) = Total annual snowfall estimated as the sum of total daily precipitation for days with an average temperature under 0 °C.

AFI (freezing degree-days) = Sum of the average daily temperatures multiplied by −1 for the period where a 14-day moving average is under 0 °C (beginning July 31 and ending July 30 of the subsequent year); daily temperatures over 0 °C are included in the summation.

ContFrz (days) = Length in days that a 30-day moving average of average daily temperature is under 0 °C (beginning July 31 and ending July 30 of the subsequent year).

FallFrStart = Julian day (j.d.) of the first day with a minimum temperature under 0 °C (beginning July 31).

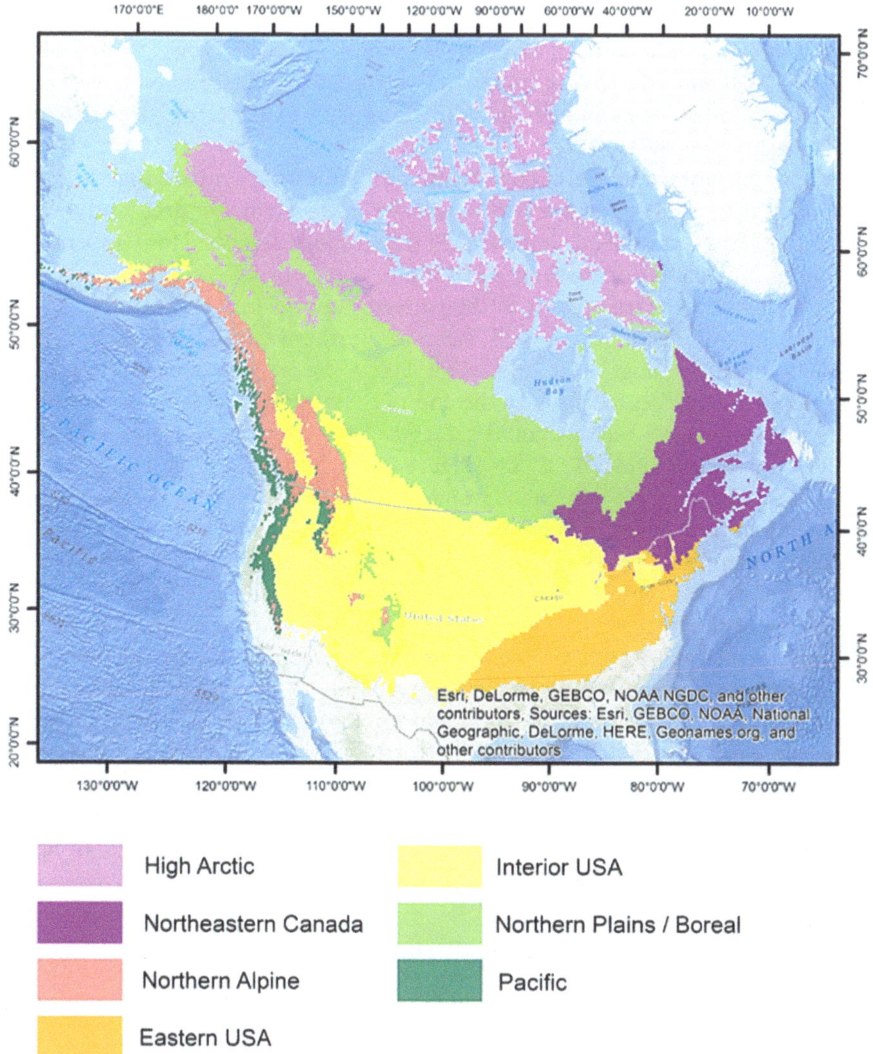

Figure 5-2. Climatic subregions over North America identified based on unsupervised k-means *clustering of 1981–2010 average winter and frost-action climate indicators.*

5.2 EXAMPLES OF CLIMATE CHANGE FOR NORTH AMERICA

5.2.1 Temperature-Related Variables

As presented in Table 5-1, the projections indicate increases in December–February average temperatures (*Tavg DJF*) for all subregions. North-south gradients, with increased warming at higher latitudes, are also observed, consistent with previous studies (e.g., Collins et al. 2013, Maloney et al. 2014). The High Arctic subregion shows the largest changes for both future horizons (+4.9 to 6.9 °C for 2041–2070 and +6.0 to 11.4 °C for 2071–2100), while the smallest are seen for the Eastern United States and Pacific subregions (+2.0 to 2.7 °C for 2041–2070 and +2.3 to 4.2 °C for 2071–2100).

An important variable related to temperature for frost-action design is the air freezing index (AFI). Positive AFI indicates lower temperatures and longer freezing seasons. Projected changes in AFI reflect those described for temperature, indicating important decreases in AFI for all subregions (Table 5-1, Figure 5-3). Latitudinal gradients are present in the pattern of projected change (Figure 5-3), with the largest decreases in AFI seen in northern latitudes and projected ensemble mean changes for the High Arctic subregion reaching between −1,312 and −2,337 degree-days by the end of the century (Table 5-1). Individual locations in northern regions could see end-of-the-century decreases exceeding −3,000 degree-days under RCP 8.5 (Figure 5-3).

The results indicate consistent trends for numerous climate indices derived from temperature values. For example, climate indices related to the timing and length of the freezing seasons, such as fall freeze-up dates (*FallFrStart*) and length of the continuous freeze (*ContFrz*), all illustrate coherent changes, indicating a shortening of the freezing season. *FallFrStart* dates are projected to occur later in the year across all subregions, from relatively modest changes in southern subregions (e.g., ensemble mean between 2 and 10 days later in the Pacific subregion, depending on future period and forcing) to relatively large changes in the north (e.g., 12 to 25 days later for the Northern Alpine subregion). Changes in *ContFrz* are projected to be even greater, as they are doubly affected by the later fall freeze dates and the earlier spring thaws. Thus, under RCP 8.5, the Northeastern Canada subregion could see ensemble mean changes in *ContFrz* of up to −50 days by the end of the century.

5.2.2 Precipitation and Snowfall Variables

December–February total precipitation (*Prec DJF*) is projected to increase for all subregions (Table 5-1), with northern regions again being those with the greatest projected ensemble mean changes. We project increases for the High Arctic subregion of between 28% and 33% for 2041–2070 and between 35% and 62% for 2071–2100. Note that this region has very dry winter conditions and typically did not receive much precipitation in the historical period (only 21 mm on average in 1981–2010), which can amplify the percentage change. Other regions show more modest ensemble mean increases, ranging from approximately 7% to 20% in 2041–2070, depending on the subregion, but possibly reaching between 10% and 36% by the end of the century. The *Prec DJF* results are subject to much greater uncertainty than the temperature projections. In some regions, model spread (indicated as $\pm 1\ \sigma$ values in Table 5-1) is almost as large as the projected ensemble mean change. For example, the 2041–2070 RCP 8.5 ensemble mean change for Northeastern Canada is +18.4%; but the standard deviation of the

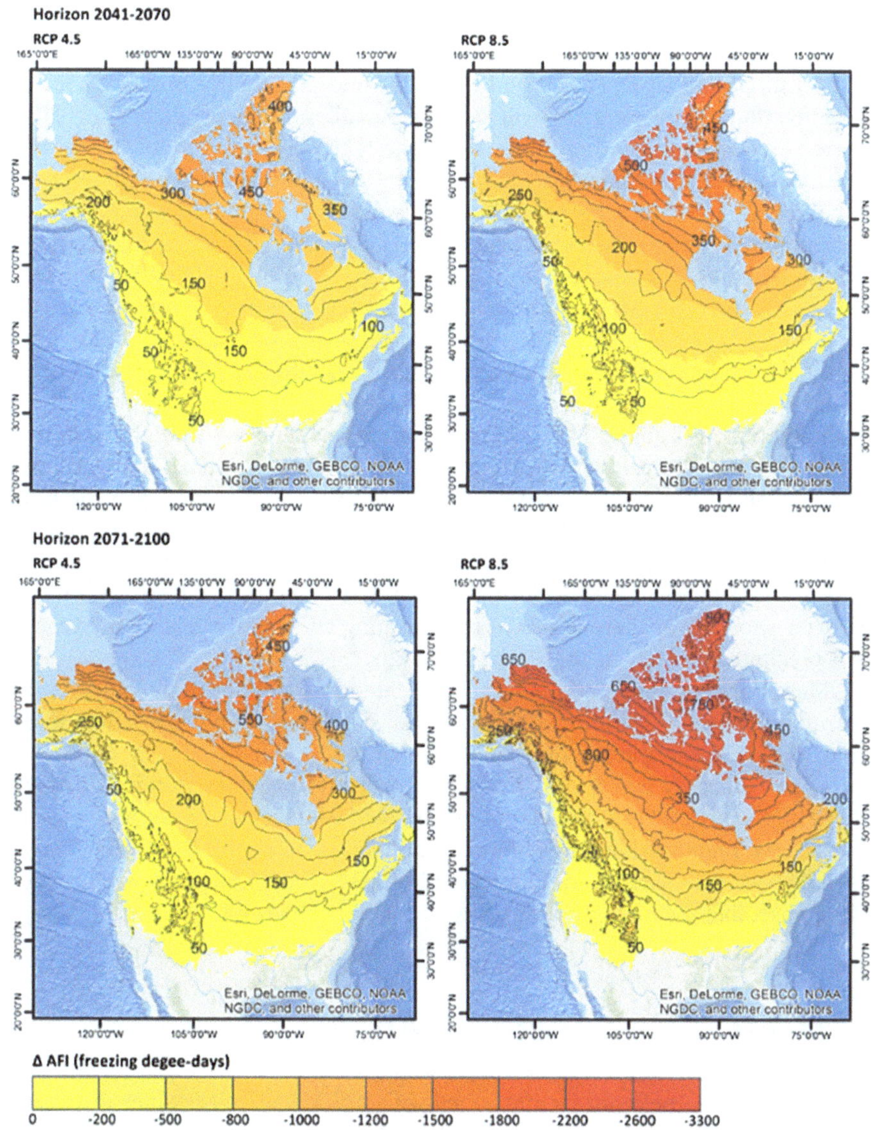

Figure 5-3. Projected changes in air freezing index (delta-AFI) across North America for 2041–2070 (upper panels) and 2071–2100 (lower panels). The maps show projected ensemble mean changes (color scale) and model uncertainty (contour lines ± 1 σ) of deltaAFI values calculated from 21 different NEX-GDDP simulations.

associated climate simulations is 12.4%. Emissions uncertainty is also present, particularly for 2071–2100, where RCP 8.5 mean changes are often twice the projected RCP4.5 value.

Despite projected increases in total precipitation (solid plus liquid phases), rising future temperatures are likely to reduce the fraction of precipitation falling as snow in December to February. Indeed, we find that the December–February snow fraction (*SnwFrac DJF*) will decrease in all subregions except the High Arctic. The subregions that show the largest change seem to be those that currently experience average December–February temperatures relatively close to the freezing point and that typically already experience a mix of liquid and solid precipitation in these months (Table 5-1). For example, the Interior USA subregion shows an observed 1981–2010 *Tavg DJF* of −3.6 °C and a *SnwFrac DJF* of 70%. In this case, even the relatively modest increases in temperature for the 2041–2070 period result in mean temperature conditions close to 0 °C, resulting in subsequent large decreases in snow fraction. A similar result occurs for the Northern Alpine subregion, as well as a more modest example with the Pacific subregion (Table 5-1). Conversely, for regions such as the High Arctic and Northern Plains/Boreal, which have very cold observed mean temperatures (−29.3 °C and −18.6 °C, respectively, for 1981–2010), we see that even the large changes in temperature result in mean conditions that are relatively far from the freezing point and delta *SnwFrac DJF* of small or no change in the future. But decreases in *SnwFrac DJF* are not necessarily indicative of less total December–February snowfall, as there is also a compensating effect of the overall increase in total precipitation seen in the *Prec DJF* results.

Many areas also receive significant amounts of snowfall outside the December–February period, and numerous subregions show projected mean reductions in total annual snowfall (*SnwFall ANN*, Table 5-1). In particular, lower *SnwFall ANN* is seen for the Eastern and Interior USA, Pacific, Northern Alpine, and Northeastern Canada subregions. Model uncertainty for these regions is relatively high, making it somewhat difficult to determine to what extent *SnwFall ANN* will be reduced. The remaining subregions show little or no change in the ensemble mean change, again with relatively high uncertainty.

5.3 RELATIONSHIP BETWEEN CLIMATE AND PAVEMENTS

Climate is one of the main factors, along with traffic, that contributes to flexible pavement deterioration (Willway et al. 2008, Transportation Association of Canada 1997, Qiao et al. 2013). For example, the effect of climate on total pavement damage has been reported as in the range of 60% to 75% (Tighe 2002) or 45% to 80% (Doré et al. 2006). As mentioned, climate-related parameters, such as temperature and precipitation, have a direct influence on pavement and soil response under frost action. For example, in permafrost areas, differences in climate result in changes of permafrost distribution and the extent of ground ice within permafrost (Transportation Association of Canada 2010). Even minor increases in heat loss at the surface of the ground produce thaw and increase active-layer thickness in permafrost regions (Hayley and McGregor 2007).

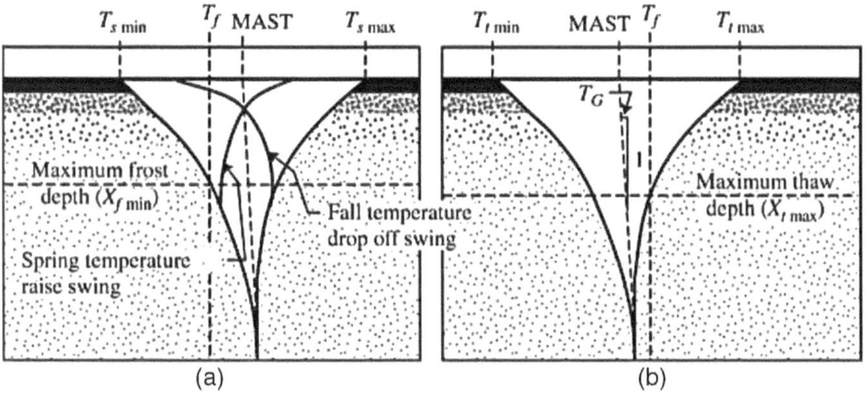

Figure 5-4. Temperature trumpet curves for (a) seasonal frost conditions and (b) permafrost conditions.

Source: Adapted from Doré and Zubeck (2009).

Note: TG = thermal gradient, MAST = Mean Annual Surface Temperature, T_s or T_t = surface temperature, T_f = freezing temperature.

As mentioned, one of the key parameters related to frost heave and ice-lens growth in soils is the thermal gradient, T_G, to which the pavement layers and soils are subjected (Andersland and Ladanyi 2004). As shown in Figure 5-4, with typical trumpet curves for (a) seasonal frost and (b) permafrost conditions, the temperatures encountered at a specific location directly affect the thermal boundaries in the pavement system. At the pavement surface, the surface temperature, T_s, which can be expressed in various ways as a function of air temperature, acts as the top thermal boundary condition. The mean annual surface temperature (MAST) describes the average annual conditions at the top of the pavement surface. Air temperature variations throughout the yearly cycle will cause temperature variations in the soil, which are gradually attenuated with depth in the pavement system (Andersland and Ladanyi 2004, Ladanyi 1996) until surface-temperature variations are completely damped. And both surface temperatures and ground temperatures are a function of climatic conditions, which are themselves changing.

During late fall, with lower temperatures induced at the pavement surface, the temperature profiles slowly shift toward the left (cold) part of the trumpet curve, and the pavement system slowly switches from summer heat-induction conditions to winter heat-extraction conditions (Doré and Zubeck 2009). In freezing conditions, this gradual shift of surface temperatures induces frost penetration in the pavement system up to a maximum frost depth, $X_{f\,min}$. Early, during freeze-up, the temperature gradients are steep, and the heat extraction rate is high; this is associated with faster frost-front penetration. A gradual decrease of the temperature gradient, with respect to time and depth, causes frost penetration rate to slow down, and frost depth reaches a maximum value. When frost penetration reaches frost-susceptible soils and when water is available, ice-lens growth will occur. This process is one of the main pavement performance issues in cold regions (Doré 1997). Thinner ice lenses will grow higher up in the pavement

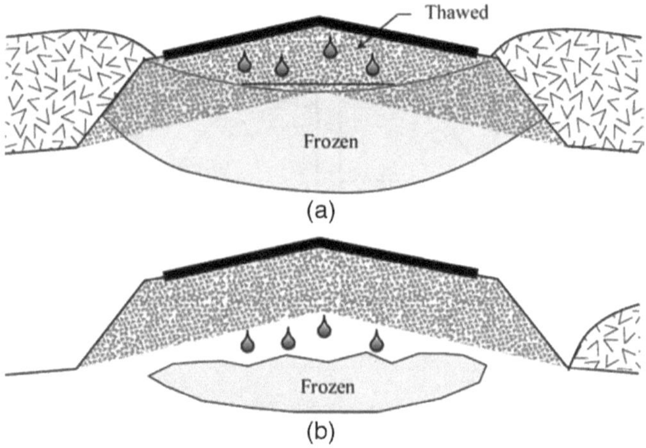

Figure 5-5. Pavement thawing in (a) early spring and winter partial thaw and (b) late spring.
Source: Adapted from Doré and Zubeck (2009).

system because of the faster heat extraction and the transient thermal conditions. Ice lenses are typically thicker with slower heat exchange, when temperature gradients are milder and near-steady-state thermal conditions are encountered. However, the formation of ice lenses is mostly marginal when the degree of saturation is below 70% (Konrad and Roy 2000). During winter, when thawing temperatures last long enough to shift the top part of the trumpet curve above freezing, the top part of the pavement thaws. This has been described as challenging performance conditions for pavement engineers (White and Coree 1990, Drolet 2015). As mentioned in this book the phenomenon is similar to early spring conditions,when the upper part of the pavement is thawed and saturated by water, made available from melting ice and snow, that is trapped between the thaw front and the asphalt layer (Figure 5-5). Significant fatigue damage typically occurs during this period (Doré and Zubeck 2009). These conditions, similar to early spring, combined with late-spring weakening associated with melting of segregation ice and high saturation of soils, are the main factors in thaw weakening of pavement structures (Doré and Zubeck 2009, Janoo and Shoop 2004, Carrera et al. 2009, Shoop et al. 2006). In permafrost environments, when thaw occurs, the excess water is gradually drained out of the ground, and consolidation leads to substantial settlement (Transportation Association of Canada 2010).

5.4 EXAMPLES OF KEY CLIMATE CHANGE ISSUES FOR PAVEMENTS

Climate change poses a long-term threat to the viability and durability of roads through incremental changes in surface conditions and use (Schweikert et al. 2014). However, while research shows an effect of climate change on performance and serviceability, these effects are small or negligible in some cases. For instance, the effect of global warming on pavement design was deemed negligible in Europe (Wistuba and Walther 2013). Climate changes are expected to influence pavement performance through various damage mechanisms, which translate into rutting, longitudinal cracking, fatigue cracking, transverse cracking, or roughness (Tighe et al. 2008, Kim et al. 2005, Graves and Mahboub 2006, Dawson 2010, Carrera et al. 2009, Thiam et al. 2013). More specifically, the expected climate changes in winter will unavoidably influence frost-action mechanisms and winter conditions. One of the most challenging issues to consider for flexible pavement performance is temperature (Byram et al. 2012), which is the most probable climate variation for the future (Emery 2010). For example, a winter temperature increase, as opposed to a summer increase, is expected for many regions in North America (IPCC 2013, Emery 2010). Also, the expected climate change will have a significant impact on the length of the freezing season (decrease) or on the number of days to reach a critical freezing index (increase) (Mills et al. 2009). The future climate will be more prone to frequent freezing and thawing during the winter (Logan et al. 2011, Emery 2010, Dawson 2010, Chaumont and Brown 2010). This is likely to change some of the critical pavement environment conditions and how the pavement system responds to frost action:

- *Higher surface temperatures in winter*. Because of the temperature increase expected for winter, shallower temperature gradients will develop. As a result, less frost penetration is expected. Corresponding frost heave and associated deterioration are therefore expected to change (Saint-Laurent 2012, Carrera et al. 2009). This change will also influence low-temperature transverse cracking of asphalt concrete surface layers (Kim and al. 2005, Dawson 2010, Hoff and Lalague 2010, Saha et al. 2014, Tighe et al. 2008) and longitudinal frost-heave cracking (Doré and Zubeck 2009). However, some research (Mills et al. 2009) has found, using Mechanistic-Empirical Pavement Design Guide (MEPDG) modelling for Canadian climatic conditions, that climate change should not have any significant effect on transverse cracking.
- *Greater number of freeze–thaw cycles*. The literature on climate change suggests that an increase in winter thawing events for some regions is likely (Dawson 2010, Carrera et al. 2009). This increase may be observed in northern regions, while southern regions may see fewer freeze–thaw cycles (Emery 2010). Freeze–thaw cycles are critical for pavement performance, as they create conditions similar to early spring thaw (Doré and Zubeck 2009, Konrad and Roy 2000, White and Coree 1990, Bilodeau et al. 2015, US Army 1985, Nix 2001, Badiane et al. 2015) and can lead to faster deterioration through rutting in the unbound layers and fatigue cracking of the asphalt concrete. Although it is difficult to assess, studies have explored the relationship between the freezing index and freeze–thaw cycles (Drolet 2015, Bilodeau et al. 2015, Jackson and Puccinelli 2006).

5.5 EFFECT OF REDUCED FROST PENETRATION

In winter, one of the main issues associated with freezing of soils is the frost heave created by the formation of ice lenses in frost subgrades. As mentioned, over a longitudinal pavement profile, frost heave is rarely uniform (Doré 1997, Djonkamla et al. 2015), and this is of great importance for pavement performance in cold regions. Uneven frost heave leads to pavement smoothness deterioration during winter (Doré 1997, Saint-Laurent 2012, Bilodeau and Doré 2013, Gustavsson et al. 1999; Konrad 1998, US Army 1985, Shuichi et al. 2002), and the corresponding seasonal variation of the international roughness index (IRI) is related to long-term deterioration rate (Figure 5-6).

Humphrey (2008) discussed a general reduction of frost-related damage due to a lower AFI in future climates. A lower AFI is likely to cause less frost penetration, given the relationship between the square root of AFI and frost depth (Doré and Zubeck 2009, Andersland and Ladanyi 2004, Aldrich and Paynter 1966). Lower temperature gradients mean slower frost penetration rate in pavement structure materials and subgrade soils, the former being mostly insensitive to frost action in normal conditions, while the latter may be quite sensitive. The shallower frost penetration due to lower AFI may also keep the frost front further away from soils with higher moisture, which can reduce frost heave significantly (Konrad and Roy 2000). But the lower thermal gradients are likely to mean slower but thicker ice-lens growth (Konrad 1987, Konrad and Morgernstern 1981). Lastly, the reduced frost penetration was found to be associated with less pavement life consumed during spring recovery of pavement structures (Orr and Irwin 2006).

The effect of reduced frost penetration and frost heave has been quantified (Drolet 2015, Bilodeau et al. 2015). Using an empirical relationship of average frost heave to IRI yearly deterioration rate (Doré 2002) as the basis of their approach,

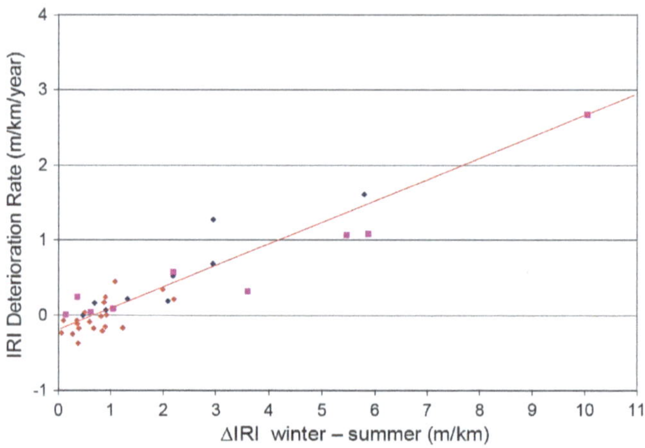

Figure 5-6. Long-term international roughness index deterioration rate with respect to seasonal variation of international roughness index.
Source: Adapted from Saint-Laurent (2012).

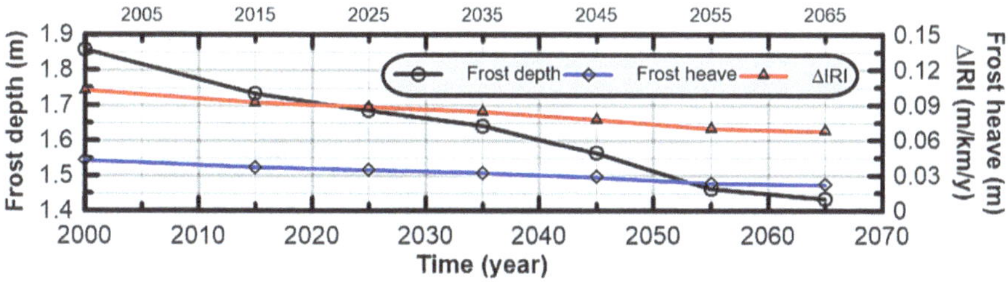

Figure 5-7. Example of the relationship between future decreasing frost depth, decreasing frost heave, and decreasing roughness yearly pavement deterioration rate for Southern Québec regional road, silty sand subgrade.
Source: Adapted from Bilodeau et al. (2015).

Bilodeau et al. (2015) projected an increase of pavement life for the 2011–2040 period, in terms of roughness for various soils and pavement functional class, of 8% to 15% for a median greenhouse gas emission increase scenario (Eastern Canada climatic conditions). This increase was quantified by using average frost-heave calculations based on the AFI (for a reference period and future periods) and the thermal balance equation (Saarelainen 1992). Figure 5-7 presents some examples of the results of Bilodeau et al. (2015).

Studies report conflicting results on the influence of rising temperatures on pavement roughness and lifespan. Saha et al. (2014) showed that, in a warming climate, IRI was unlikely to be the most affected performance parameter for flexible pavements in Canada. Some studies have found that roughness is mostly insensitive to climate changes (Kim et al. 2005). But Hoff and Lalague (2010) showed a minimal effect of climate change on IRI of about −2.7% to 8.3%, with typical increases in the range of 1% to 2%, for various pavement types and locations. And using calculations with the MEPDG, Mills et al. (2009) showed a decrease of pavement life, expressed with IRI, when climate changes were considered for flexible pavement design.

5.6 INCREASING NUMBER OF THAWING EVENTS IN THE WINTER PERIOD

The higher winter temperatures and the corresponding lower AFI are likely to alter the frequency of freeze–thaw cycle events (Emery 2010, Dawson 2010, Carrera et al. 2009, Humphrey 2008). The number of freeze–thaw cycles can be related to some extent to AFI, especially for lower values of the freezing index (Figure 5-8). A similar observation was reported by Drolet (2015). The higher number of winter partial thaws in some regions may have a significant impact on the flexible pavement deterioration rate. As the number of days in a yearly cycle during which pavement structures are subjected to spring thaw conditions is

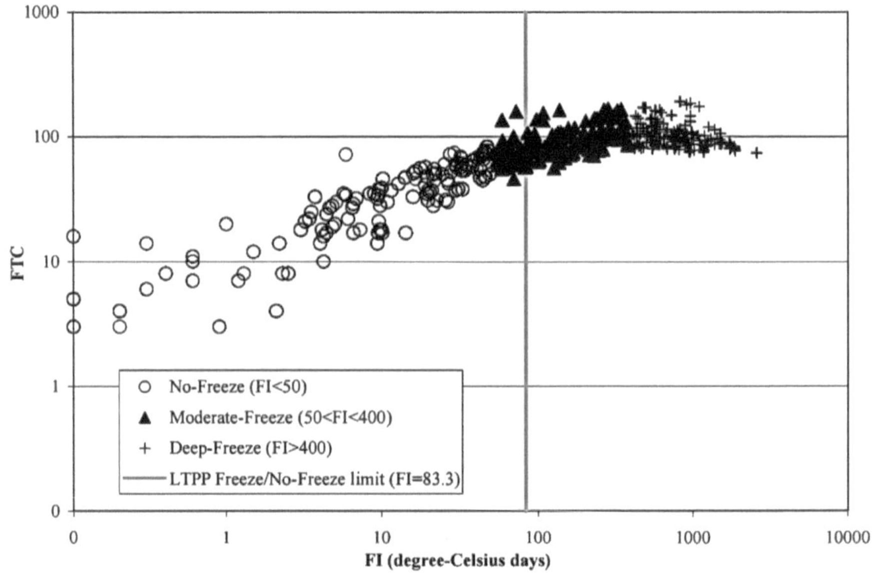

Figure 5-8. Freeze–thaw cycles (FTC) with respect to the freezing index (FI).
Source: Adapted from Jackson and Puccinelli (2006).

proportional to the amount of damage occurring during the spring period (Orr and Irwin 2006), the increase of winter thawing events is of great concern. For example, Doré and Savard (1998) observed two flexible pavement test sections and concluded that more than 90% of total fatigue damage occurred during partial winter thaw and spring thaw periods. White and Coree (1990) reported that 60% of the failures seen in the AASHTO tests occurred during thawing. Saint-Laurent and Corbin (2003) found that 30% to 70% of pavement damage occurs during the thawing period. Jackson and Puccinelli (2006) reported that the effect of the climatic zone and of the number of freeze–thaw cycles was significant for fatigue cracking and rutting, among other things, with 95% confidence.

Figure 5-9 presents examples of quantification of the effect of freeze–thaw cycles on pavement performance. Jackson and Puccinelli (2006) focused on various climatic regions and their respective variations in terms of freeze–thaw cycles to assess climatic impacts on pavement performance. Regions with a high frequency of freeze–thaw cycles, associated with moderate freeze conditions, presented the highest mean prediction for rutting based on testing of a flexible pavement with a dense graded aggregate base on a fine subgrade. Compared to no-freeze conditions, the mean prediction for rutting could be 1.64 times that in high freeze–thaw cycle climatic zones after 20 years. Bilodeau et al. (2015) focused on the effect of winter partial thaws on pavement performance. Using cyclic triaxial testing, the permanent deformation was measured for multiple repetitions of freeze–thaw cycles and for test conditions representative of the ones

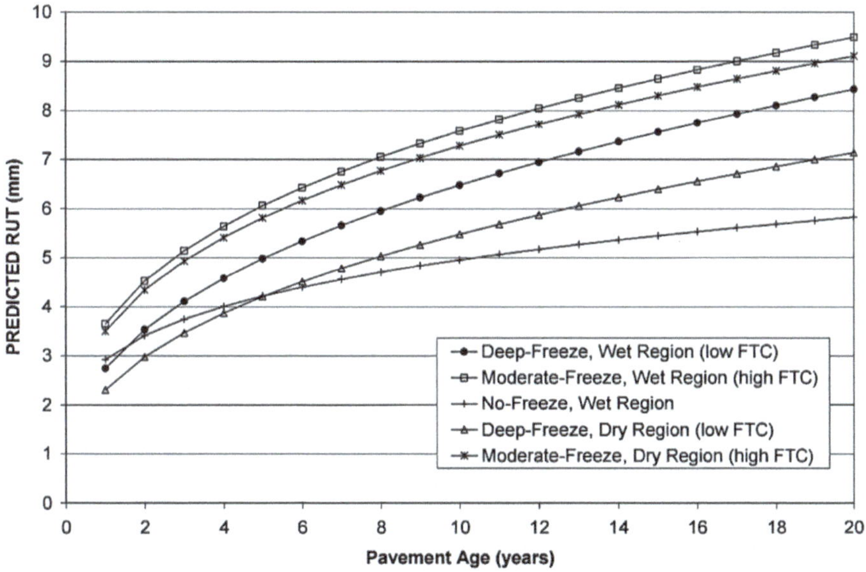

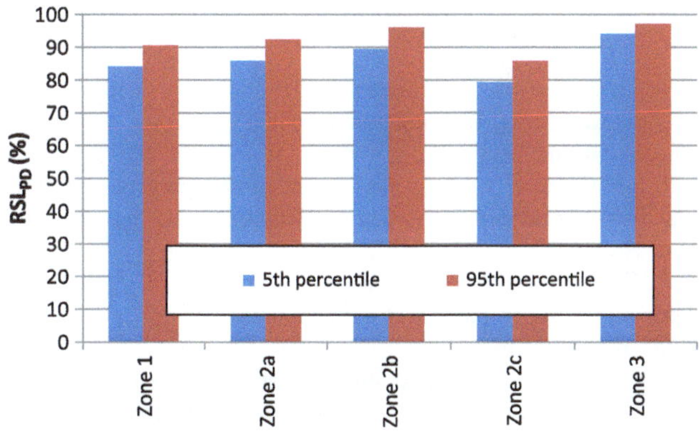

Figure 5-9. Examples of the effect of freeze–thaw cycles (FTC) on pavement performance: (top) rutting for different climatic zones; (bottom) relative service life for permanent deformation (RSL_{PD}) in the granular base with consideration of increasing winter partial thaw events for different climatic zones for plant hardiness in Québec.

Source: Top, adapted from Jackson and Puccinelli (2006); bottom, adapted from Bilodeau et al. (2015).

encountered in the granular base during winter thawing events. Using a seasonal damage analysis for permanent deformation and evaluating for different greenhouse gas increase scenarios, the average decrease of service life (relative service life for permanent deformation, in the granular base) was found to be about 10% (varying between 4% and 17%, depending on the climatic zone).

5.7 CONCLUSIONS

This chapter has described some key aspects of climate change, such as the main mechanisms involved, the modelling of future climate, and the expected trends in climate-related parameters. The evolving climate trends, with focus on the freezing period, were shown using some of the tools and hypotheses that are of general use within the scientific community. We discussed the parameters associated with temperature and precipitation with respect to their expected future evolution in North America. On the basis of those findings, we reviewed some of the key climate parameters associated with flexible pavement performance in cold regions and their effects. In the evolving climate, the changing frost heave response of the pavement system and the changing number of freezing and thawing events are among the parameters that will influence pavement performance.

References

Aldrich, H. P., and H. M. Paynter. 1966. *Depth of frost penetration in non-uniform soil.* CRREL Special Rep. No. 104. Hanover, NH: US Army Cold Regions Research and Engineering Laboratory.

Andersland, O. B., and B. Ladanyi. 2004. *An introduction to frozen ground engineering.* Hoboken, NJ: Wiley.

Badiane, M., J. Yi, G. Doré, J.-P. Bilodeau, and F. Prophète. 2015. "Monitoring of flexible pavement structure during freezing and thawing." In *Proc., Int. Conf. on Cold Regions Engineering*, Salt Lake City.

Bilodeau, J.-P., and G. Doré. 2013. "A new pavement design procedure for frost protection in seasonal frost areas." In *Proc., Bearing Capacity of Roads Railways and Airfields*, Trondheim, Norway.

Bilodeau, J.-P., F. P. Drolet, G. Doré, and M.-F. Sotile. 2015. "Effect of climate changes expected during winter on pavement performance." In *Proc., Int. Conf. on Cold Regions Engineering*, Salt Lake City.

Byram, D., D. X. Xiao, K. C. P. Wang, and K. Hall. 2012. "Sensitivity analysis of climatic influence on MEPDG flexible pavement performance prediction." In *Proc., 91st Annual Meeting of the Transportation Research Board*, Washington, DC.

Carrera, A., A. Dawson, and J. Steger. 2009. *State of the art of likely effect of climate on current roads.* Rep. No. 1. Nottingham, UK: Univ. of Nottingham.

Charron, I. 2014. *A guidebook on climate scenarios: Using climate information to guide adaptation research and decisions.* Montreal: Ouranos.

Chaumont, R. D., and R. D. Brown. 2010. "Analyse de simulations régionales du climat et d'indices climatiques associés au transport routier dans le Sud du Québec." *Revue Routes-Road* **345**: 78–83.

Collins, M., R. Knutti, J. Arblaster, J.-L. Dufresne, T. Fichefet, P. Friedlingstein, et al. 2013. "Long-term climate change: Projections, commitments and irreversibility." Chap. 12 in *Climate change 2013: The physical science basis. Chapter contribution of working group i to the fifth assessment report of the intergovernmental panel on climate change*, T. F. Stocker, D. Qin, G.-K. Plattner, M. Tignor, S. K. Allen, J. Boschung, et al., eds., 1029–1136. Cambridge, UK: Cambridge University Press.

Dawson, A. 2010. *Pavement performance and remediation requirements following climate change.* Era-Net Road. Nottingham, UK: Univ. of Nottingham.

Djonkamla, Y., G. Doré, and J.-P. Bilodeau. 2015. "Développement d'un modèle mécaniste-empirique de prédiction de l'uni des chaussées flexibles." In *Proc., 68e Conférence Canadienne de Géotechnique et 7e Conférence Canadienne sur le Pergélisol.* Québec City, Québec.

Doré, G. 1997. "Détérioration des chaussées en conditions de gel: une nouvelle approche prévisionnelle." Ph.D. dissertation, Département de Génie Civil, Université Laval.

Doré, G. 2002. *Validation des modèles de détérioration de la chaussée soumise au soulèvement par le gel.* Rapport de recherche GCT-02-07. Québec: Département de génie civil, Université Laval.

Doré, G., P. Drouin, P. Pierre, P. Desrochers, and P. Ullidtz. 2006. "Estimation of the relationships of flexible pavement deterioration to traffic and weather in Canada." In *Proc., 10th Int. Conf. on Asphalt Pavements (CD-ROM)*, Québec.

Doré, G., and Y. Savard. 1998. *Analysis of seasonal pavement deterioration.* Washington, DC: Transportation Research Board, National Academies.

Doré, G., and H. Zubeck. 2009. *Cold regions pavement engineering.* New York: McGraw-Hill.

Drolet, F. P. 2015. "Effets des changements climatiques sur la performance à long terme des chaussées souples au Québec: Volet 2-Effet de l'augmentation de la température en hiver et d'une hausse du nombre d'épisode de redoux hivernaux." M.S. thesis, Département de Génie Civil, Université Laval.

Emery, J. 2010. "Climate change impacts on asphalt pavements: Global perspective, adaptation and opportunities." In Proc., *Canadian User Producer Group of Asphalt.*

Graves, R. C., and K. C. Mahboub. 2006. "Pilot study in sampling-based sensitivity analysis of NCHRP design guide for flexible pavements." *Transp. Res. Rec.* **1947** (1): 123–135.

Gustavsson, H., E. Slunga, and S. Saarelainen. 1999. "Finnish frost design concept of roads, geotechnical engineering for transportation infrastructure." In *Proc., 12th European Conf. on Soil Mechanics and Geotechnical Engineering*, Amsterdam, The Netherlands, 1359–1366.

Hayley, D. W., and R. V. McGregor. 2007. "The importance of permafrost, ice and seasonally frozen ground to road systems in Canada." In *Proc., 23rd PIARC World Road Congress*, Paris.

Hoff, I., and A. Lalague. 2010. *Analysis of pavement structural performance for future climate.* Rep. No. 7, Era-Net Road, Project P2R2C2. Nottingham, UK: Univ. of Nottingham.

Humphrey, N. 2008. "Potential impacts of climate change on U.S. transportation." Special Report 290. Washington, DC: National Academies Press. https://www.doi.org/10.17226/12179. Accessed January 30, 2020.

IPCC (Intergovernmental Panel on Climate Change). 2013. *Climate change 2013: The physical science basis. Chapter contribution of working group I to the Fifth assessment report of the intergovernmental panel on climate change*, T. F. Stocker, D. Qin, G.-K. Plattner, M. Tignor, S. K. Allen, J. Boschung, et al., eds. Cambridge, UK: Cambridge Univ. Press.

Jackson, N., and J. Puccinelli. 2006. *Long-term pavement performance (LTPP) Data analysis support: National pooled fund study tpf-5(013), effects of multiple freeze cycles and deep frost penetration on pavement performance and cost.* Rep. No. FHWA-HRT-06-121. Washington, DC: Federal Highway Administration, US Dept. of Transportation.

Janoo, V., and S. Shoop. 2004. "Influence of spring thaw on pavement rutting." In *Proc., 6th Int. Symp. on Pavements Unbound (UNBAR 6)*, Nottingham, UK, 115–124.

Kim, S., H. Ceylan, and M. Heitzman. 2005. "Sensitivity study of design input parameters for two flexible pavement systems using the mechanistic empirical pavement design guide." In *Proc., Mid-Continent Transportation Research Symp.* Ames, Iowa.

Konrad, J.-M. 1987. "The influence of heat extraction rate in freezing soils." *Cold Reg. Sci. Technol.* **14** (2): 129–137.

Konrad, J.-M. 1998. "Méthodologie pour la conception de chaussées urbaines sur sols gélifs." *Revue Géoglobe* (Autumn): 12–15.

Konrad, J.-M., and N. R. Morgenstern. 1981. "The segregation potential of a freezing soil." *Can. Geotech. J.* **18** (4): 482–491.

Konrad, J.-M., and M. Roy. 2000. "Flexible pavements in cold regions: A geotechnical perspective." *Can. Geotech. J.* **37** (3): 689–699.

Ladanyi, B. 1996. *Conception et réhabilitation des infrastructures de transports en régions nordiques.* Québec: Government of Québec.

Logan, T., I. Charron, D. Chaumont, and D. Houle. 2011. *Atlas de scénarios climatiques pour la forêt québécoise.* Québec: et de la Faune de Québec, Government Report for the Province of Québec.

Maloney, E. D., S. J. Camargo, E. Chang, B. Colle, R. Fu, K. L. Geil, et al. 2014. "North American Climate in CMIP5 experiments. Part III: Assessment of twenty-first-century projections." *J. Clim.* **27** (6): 2230–2270.

Mills, B., S. L. Tighe, J. Andrey, J. Smith, and K. Huen. 2009. "Climate change implications for flexible pavement design and performance in southern Canada." *J. Transp. Eng.* **135** (10): 773–782.

Moss, R. H., J. A. Edmonds, K. A. Hibbard, M. R. Manning, S. K. Rose, and D. P. van Vuuren. 2010. "The next generation of scenarios for climate change research and assessment." *Nature* **463** (7282): 747–756.

Nix, F. 2001. *Weight-distance taxes.* Ottawa, ON: Canadian Truck Alliance.

NOAA (National Oceanic and Atmospheric Administration) National Centers for Environmental Information. 2015. "State of the climate: Global analysis for August 2015." Accessed November 19, 2019. http://www.ncdc.noaa.gov/sotc/global/201508.

Orr, D. P., and L. H. Irwin. 2006. *Seasonal variations of in situ material properties in New York State.* Cornell University Local Roads Program Rep. No. 06-6, New York State DOT Research Project C-01-54. Ithaca, NY: Cornell Univ.

Qiao, Y., G. W. Flintsch, A. Dawson, and T. Parry. 2013. "Examining effects of climatic factors on flexible pavement performance and service life." *Transp. Res. Rec.* **2349** (1): 100–107.

Saarelainen, S. 1992. *Modelling frost heaving and frost penetration in soils at some observation sites in Finland: The SSR model.* Espoo, Finland: Technical Research Center of Finland.

Saha, J., S. Nassiri, A. Bayat, and H. Soleymani. 2014. "Evaluation of the effects of Canadian climate conditions on the MEPDG predictions for flexible pavement performance." *Int. J. Pavement Eng.* **15** (5): 392–401.

Saint-Laurent, D. 2012. "Routine mechanistic pavement design against frost heave." In *Cold regions engineering 2012: Sustainable infrastructure development in a changing cold environment*, 144–154. Reston, VA: ASCE.

Saint-Laurent, D., and G. Corbin. 2003. "L'impact des restrictions de charge en période de dégel." *Innovation Transp.* **18**: 25–31.

Schweikert, A., P. Chinowsky, X. Espinet, and M. Tarbert. 2014. "Climate change and infrastructure impacts: Comparing the impact on roads in ten countries through 2100." *Procedia Eng.* **78**: 306–316.

Sheffield, J., G. Goteti, and E. F. Wood. 2006. "Development of a 50-year high-resolution global dataset of meteorological forcings for land surface modeling." *J. Clim.* **19** (13): 3088–3111.

Shoop, S., R. Haehnel, V. Janoo, D. Harjes, and R. Liston. 2006. "Seasonal deterioration of unsurfaced roads." *J. Geotech. Environ. Eng.* **132** (7): 852–860.

Shuichi, K., K. Atsushi, K. Mitsuo, K. Akira, and H. Kenji. 2002. "Effect of frost heave on the longitudinal profile of asphalt pavements in cold regions." In Vol. **1** of *Proc., 9th Int. Conf. on Asphalt Pavements*, Copenhagen, Denmark, 9–1.

Taylor, K. E., R. J. Stouffer, and G. A. Meehl. 2012. "An overview of CMIP5 and the experiment design." *Bull. Am. Meteorol. Soc.* **93** (4): 485–498.

Thiam, P. M., G. Doré, and J. P. Bilodeau. 2013. "Effect of the future increases of precipitation on the long-term performance of roads." In *Proc., 9th Int. Conf. on the Bearing Capacity of Roads Railways and Airfields*, Trondheim, Norway.

Tighe, S. L. 2002. "Evaluation of subgrade and climatic zone influences on pavement performance in the Canadian Strategic Highway Program's (C-SHRP) Long-Term Pavement Performance (LTPP) study." *Can. Geotech. J.* **39** (2): 377–387.

Tighe, S. L., J. Smith, B. Mills, and J. Andrey. 2008. "Evaluating climate change impact on low-volume roads in southern Canada." *Transp. Res. Rec.* **2053** (1): 9–16.

Transportation Association of Canada. 1997. *Pavement design and management guide*, R. Haas, ed. Waterloo, Canada: Transportation Association of Canada.

Transportation Association of Canada. 2010. *Guidelines for development and management of transportation infrastructure in permafrost regions.* Ottawa: Transportation Association of Canada.

US Army. 1985. *Pavement design for seasonal frost conditions.* ARMY TM 5-818-2. Washington, DC: Dept. of the Army.

White, T. D., and B. J. Coree. 1990. "Threshold pavement thickness to survive spring thaw." In *Proc., 3rd Int. Conf. on Bearing Capacity to Survive Spring Thaw*, Trondheim, Norway.

Willway, T., L. Baldachin, S. Reeves, M. Harding, M. McHale, and M. Nunn. 2008. *The effects of climate change on highway pavements and how to minimise them.* Technical Rep. No. PPR184. Berkshire, UK: Transport Research Laboratory.

Wistuba, M. P., and A. Walther. 2013. "Consideration of climate change in the mechanistic pavement design." Supplement, *Road Mater. Pavement Des.* **14** (S1): 227–241.

PART II

Mitigation of Frost Action in Pavements

Introduction

Guy Doré

Frost action detrimentally affects pavement structures in two ways: pavement heaving by segregation freezing in frost-susceptible soils, and loss of bearing capacity of the pavement structure once the ice contained in the base course, subbase, and subgrade soil thaws. There are several mitigation measures used to attenuate the effects of these phenomena, all based on empirical procedures (Dysli 1991). However, the use of pavement design approaches based on rational and physical models and on analytical and numerical techniques is increasing.

Most of the agencies managing road networks in northern Europe, Asia, and North America use design criteria related to frost action. But surprisingly, several agencies do not use specific design procedures to mitigate frost action, despite frost action significantly affecting their road network.

The following chapters synthesize the principles, criteria, and methods used for frost protection of pavement and other structures in cold climates.

Reference

Dysli, M. 1991. *Le gel: Et son action sur les sols et les foundations.* Lausanne, Switzerland: Presses polytechniques et universitaires romandes.

CHAPTER 6
The Design Principles

Guy Doré

There are a wide range of approaches to structural design in freezing conditions. Most use one or both of these two main strategies:

- Design the structural components of the pavement and verify how the structure will be affected by frost heave or frost penetration. If frost heave or frost penetration exceeds allowable values based on criteria adopted by the highway agencies, the structure must be improved.
- Design the structural components of the pavement without considering frost action but verify the design for adequate bearing capacity during spring thaw. The structure must be improved if it cannot sufficiently withstand traffic load during the thaw season.

The designer must first establish which mechanism, frost heave or thaw weakening, is likely to have the greatest impact on a specific pavement. The designer may then choose to focus on the control of frost heave or only consider the loss of support capacity during the thaw season.

In the first case, the strategy will involve controlling one of the three factors essential to the segregation freezing process: the frost susceptibility of subgrade soil; temperatures below the freezing point for a significant period of time, allowing frost to penetrate the pavement structure and the subgrade soil; and the availability of water in the soil near the frost front. In the second case, the designer will focus on the bearing-capacity loss during spring thaw, and seasonal (mainly spring) variation of mechanical properties of subgrade soil and pavement materials will be a determining factor in the design. The designer can also try to account for both phenomena, knowing that control of the segregation freezing process will help mitigate the loss of bearing capacity due to thaw.

The designer can also choose between two levels of pavement protection. One is to attenuate the effects of freeze–thaw cycles and to adapt the pavement structure to resist these attenuated freeze–thaw effects. This is generally referred to as a partial protection approach. Partial protection implies accepting a seasonal loss of functional or structural capacity of the pavement associated with frost action. In partial protection, the pavement structure can be made stiffer or more flexible as a function of the significance of the frost heave during winter or the bearing-capacity loss due to thaw. Consequently, the objectives of the design, lifetime expectancy, and service level offered to users must be adjusted.

The other option is the total protection approach: trying to neutralize the damaging effects of freeze–thaw cycles. This strategy focuses on controlling one of the three factors essential for segregation freezing. Total protection aims to maintain the functional performance and structural capacity throughout the year. The design must be made in consideration of the worst annual conditions (freeze or thaw), which generally leads to systematically overdesigned pavements. In all cases, it is beneficial for the pavement to be thicker to reduce frost action in the subgrade soils, to better distribute strains, and to absorb deformations. Table 6-1 summarizes the design strategies, including some examples of applicable techniques.

Furthermore, the designer may choose to intervene locally or over an entire project. For example, transitions or frost tapers are often used in locally severe cases where there is contact between frost-susceptible and non-frost-susceptible soil. General protection measures will be adopted in cases where pavements are underlain by highly frost-susceptible soils. The choices depend on local conditions and on the classification of the road considered.

Indicators are measurable values or indices related to the response of the pavement structure to climatic factors in a given geological and hydrogeological context. They can be compared to a reference value, or criterion, linked to the desired performance of the pavement. This type of design criterion is based on local experience, suggesting that pavements that do not meet the criterion will not have acceptable performance. Frost depth is an indicator frequently used in verification methods. Many methods use an empirical criterion related to this indicator to verify the adequacy of the pavement structure. For example, the total thickness of the pavement structure should correspond to 70% of the expected frost depth (Kono et al. 1973). Some agencies have raised the sophistication of this simple criterion by adding soil and road classification correction factors to take into consideration different levels of frost susceptibility of subgrade soils and different levels of risk tolerance depending on the functional classification of the road (St-Laurent 2006).

Indicators can also be used as weighting factors for other design variables. For example, the original AASHTO method used a “regional factor” to weight the effect of traffic as a function of frost conditions (Johnson 1973). Lastly, the indicator can be directly used as a design factor. This is particularly the case in the Finnish design method (Jämsä and Orama 1990, Dysli 1991), which uses a spring soil support modulus as a basis for pavement structural design. But this type of approach can be overly conservative and lead to overdesign of the pavement structure.

There is a large range of types and qualities of indicators and criteria for frost verification of pavement structures. Table 6-2 shows a tentative classification of these indicators according to their design application and quality. Indicators of the first order are ones that constitute a direct measure of pavement response to frost

Table 6-1. Structural Pavement Design Strategies to Counter the Effects of Freezing.

Mechanism considered	*Approach*	*Examples of applicable mitigation approaches*
Frost heaving	Structural adaptation	• Thicken the subbase (to reduce frost action in the subgrade soil). • Improve flexibility of the asphalt layer (to resist frost-related cracking).
	Neutralization of the mechanism	• Add pavement drainage (only deep drainage will have a significant effect). • Add pavement insulation. • Homogenize subgrade soils to reduce the risk of differential heaving. • Remove frost-susceptible soil and replace with granular material (to the depth of frost penetration). • Neutralize the frost susceptibility of subgrade soils with chemical treatments.
Bearing capacity	Structural adaptation	• Stiffen the structure (stabilize base materials and/or thicken the asphalt concrete layer). • Thicken the granular layers for better load distribution.
	Neutralization of the mechanism	• Add pavement drainage (only deep drainage will have a significant effect). • Use granular materials with low susceptibility to water and frost action. • Add pavement insulation. • Homogenize subgrade soils to reduce the risk of differential heaving and frost heaving. • Remove frost-susceptible soil and replace with granular material (to the depth of frost penetration) • Neutralize the frost susceptibility of subgrade soils with chemical treatments.

Table 6-2. Indicators and Criteria Used in the Verification of Permafrost Structures.

Considered mechanism	*Indicator quality*	*Indicator*	*Criteria used (admissible values)*
Frost heaving	First order	Heave	40–200 mm, depending on road classification (Finland)[1] 50–80 mm, depending on road classification (Québec) 50–300 mm, depending on road classification (Sweden)[2]
		Change of slope	4% to 15%, according to the road type (Finland)[12]
		Loss of serviceability	Δ PSI (AASHTO)[3] Δ RCI (Ontario)[2]
Loss of bearing capacity during thaw	First order	Stress Strains Deflections	Resilient modulus and damages integrated over the year (AASHTO)[3] Springtime bearing capacity (Finland)[1,4] Based on representative subgrade resilient modulus (Ontario)[2] or based on saturated CBR (Saskatchewan)[5] Note: these values are directly used in response and damage models based on fatigue cracking, rutting, or deflection.

Heaving and loss of bearing capacity	Second order	Frost penetration Allowable freezing index	$D = 0.4$ to $0.6 \times X_{30}$ (Switzerland)[1] $D = 0.8 \times X$ (CRREL)[6] $D = 0.7 \times X$ (Japan)[7,8] Allowable FI at the subgrade line (France)[1]
Heaving and loss of bearing capacity	Third order	Freezing index Frost sensitivity of the soil Regional factor	D = function of FI, SG, and road classification (Québec)[9] D = function of FI_x and SG (Norway)[1] $0 < SB < 600$ mm, according to the engineer's judgment (New Hampshire and New York)[5]

Source: Doré (1997).

Notes:

[1]Dysli (1991)
[2]Ontario Ministry of Transportation and Communications (1990)
[3]AASHTO (1993)
[4]Jämsä and Orama (1990)
[5]Johnson (1973)
[6]Ladanyi (1996)
[7]Kono et al. (1973)
[8]Kubo and Takeichi (1992)
[9]Transports Québec (1995)

B = Subbase
CBR = California bearing ratio
D = Total thickness
FI = Freezing index
PSI = Pavement serviceability index
RCI = Ride comfort index
SG = Frost susceptibility of subgrade soil
X = Frost depth

action. Second-order indicators are an indirect indication of the effect of frost or climate. Such values are specific to the structure considered. Third-order indicators are indicators of the effect of climate on the pavement but are not specific to a given structure. The table also includes some criteria used in the methods summarized in the next chapters.

References

AASHTO. 1993. *AASHTO guide for the design of pavement structures*. Washington, DC: AASHTO.

Doré, G. 1997. "Deterioration des chausses en conditions de gel: une nouvelle approche previsoinelle." Ph.D. thesis, Département de Génie Civil, Université Laval.

Dysli, M. 1991. *Le gel: Et son action sur les sols et les fondations*. Lausanne, Switzerland: Presses polytechniques et universitaires romandes.

Jämsä, H., and R. Orama. 1990. "The measurement of seasonal subgrade properties under existing pavements and the Finnish pavement design method." In *Proc., 69th Annual Meeting TRB Preprint 890619*, Washington, DC.

Johnson, T. C. 1973. "Calcul des chaussées dans les régions de gel saisonnier d'Amérique du Nord." *Compte rendu du symposium sur l'action du gel sur les chaussées*, Oslo, Norway, 124–131.

Kono, F., T. Takahashi, H. Kubo, and H. Hiraoka. 1973. "Pavement design in cold areas." In *Proc., Symp. of Frost Action on Roads*, 138–141. Oslo, Norway.

Kubo, H., and K. Takeichi. 1992. "Seasonal nondestructive evaluation of frost-heave prevention layers in asphalt pavements." *Transp. Res. Rec.* **1362**: 90–94.

Ladanyi, B. 1996. *La conception et la réhabilitation des infrastructures de transport en régions nordiques*. Direction des Communications, RTQ-94-07. Montréal: Québec Ministry of Transportation.

Ontario Ministry of Transportation and Communications. 1990. *Pavement design and rehabilitation manual*. SDO-90-01. Toronto, Canada: Surveys and Design Office.

St-Laurent, D. 2006. *Chaussée 2: Logiciel de dimensionnement des chaussées souples. Guide de l'utilisateur*. Québec: Québec Ministry of Transportation.

Transports Québec. 1995. *Normes: Ouvrages routiers, Tome I: Conception routière Ministère des Transports du Québec*. Québec: Les publications du Québec.

CHAPTER 7

Pavement Design Methods in Cold Climates

Guy Doré

This section describes several frost design or verification methods and illustrates several approaches to pavement design in frost conditions. It summarizes for the United States methods developed by AASHTO and the US Army Corps of Engineers Cold Regions Research and Engineering Laboratory (CRREL), and for Canada the methods from the Ontario Ministry of Transportation and the Québec Ministry of Transportation. For Europe, it describes methods used in France and the Nordic countries. Lastly, the Doré and Zubeck method is presented.

7.1 METHODS USED IN THE UNITED STATES

The CRREL method (Departments of the Army and Air Force 1985, Berg and Johnson 1983, Ladanyi 1996) is essentially an empirical method that includes two specific verification procedures:

1. Take into consideration the reduction of pavement bearing capacity during spring thaw (reduced subgrade strength, RSS).
2. Control differential frost heaving based on the control of frost penetration in the subgrade soil (limited subgrade frost penetration, LSFP).

Factors this method considers are the average freezing index of the three coldest winters in the last 30 years or the coldest in the last 10 years, the annual average temperature, the duration of the freezing season, the dry density of pavement materials, the water content, and the granular layer thickness. The RSS procedure uses an effective annual California bearing ratio determined by the correlation with the frost susceptibility of the subgrade soil. The total thickness of the pavement is determined empirically in the LSFP procedure using an equation or the design chart replicated in Figure 7-1. The method includes a recommendation to homogenize the subgrade soil to a depth equal to two-thirds of the frost depth or 600 mm, whichever is smaller (Berg and Johnson 1983), although

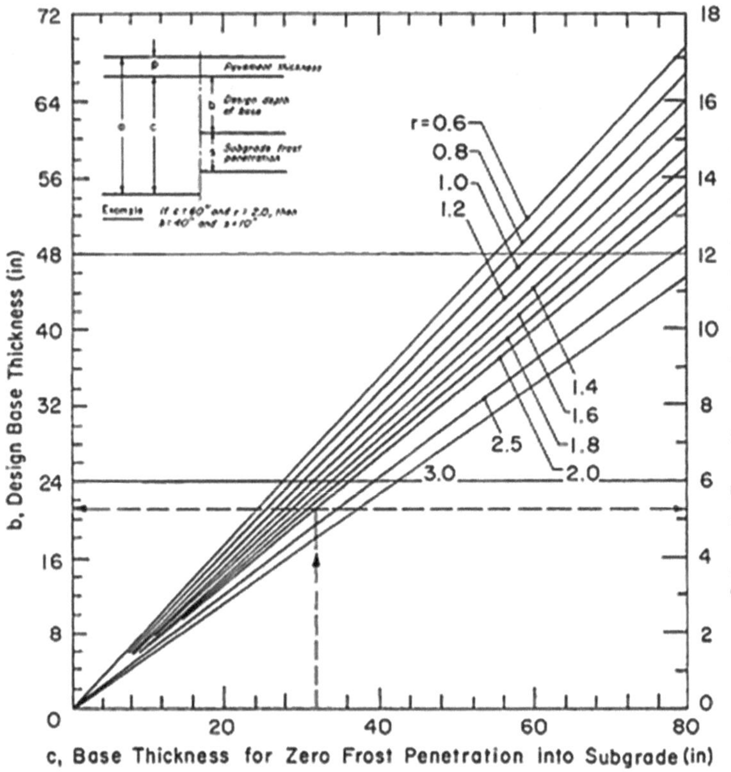

Figure 7-1. Design depth of non-frost-susceptible base for limited subgrade frost penetration.
Source: Berg and Johnson (1983).
Notes:
a = Combined thickness of pavement and nonfrost-susceptible base for zero frost penetration into the subgrade
$c = a - p$
w_b = Water content of base
w_s = Water content of subgrade
$r = w_s/w_b$, not to exceed 2.0 for airfields and 3.0 for other pavements

homogenization to this extent is dependent on the non-uniformity of the subgrade. (Robert A. Eaton, personal communication, 2016).

The AASHTO pavement design method (AASHTO 1993 and earlier versions, Forsyth 1993) is widely used in North America. The method is also empirical and is based on the largest pavement experiment ever undertaken. The main interest in this method lies in its wide acceptance. In the AASHTO 1993 version, the method proposes design tools addressing both the loss of springtime bearing capacity and the risk associated with frost heave. The method is based on the calculation of loss of serviceability (ΔPSI) by combining the effects of traffic and climate. (Note: The pavement serviceability index or PSI is an indicator of the service level a pavement offers. Developed in the early 1960s, within the AASHTO test period, it includes several parameters, of which the most important are surface degradation and ride quality; see Haas et al. 1994, Transportation Association of Canada 1997).

The AASHTO method accounts for seasonal climate variation in PSI with three parameters: roadbed (subgrade) bearing capacity, subgrade swelling, and frost heave. Subgrade bearing capacity and frost heave are of the most concern in cold regions. Variation in subgrade bearing capacity is addressed by the calculation of an effective resilient modulus for the subgrade soil based on the relative damage caused by traffic during each season, averaged over a year. The method is based on an empirical model relating the resilient modulus of subgrade soil, determined from laboratory tests performed on representative samples or by correlations between soil properties (clay content, moisture, plastic index, etc.) and relative pavement damage, taking into consideration variation in bearing capacity throughout the year. This accounts for both strength gains during winter freeze and strength loss during spring thaw. An estimated subgrade soil resilient modulus value is determined for either two-week or one-month periods throughout the year. These values are then used to estimate relative pavement damage for each period. The average of these seasonal relative damage values, the annual relative damage, is in turn used in the model to recalculate an effective resilient modulus of subgrade soil for the year. Figure 7-2 shows the AASHTO chart, with examples for the calculation of the roadbed soil effective resilient modulus.

The AASHTO method also proposes an empirical model to estimate the loss of serviceability related to frost heave. This approach, illustrated in Figure 7-3(a), is based on two design charts. The first chart, Figure 7-3(b), allows calculating the maximum loss of serviceability (ΔPSI_{max}) attributable to frost action according to the frost-front penetration under the surface of the pavement and the quality of the drainage. The ΔPSI_{max} is then adjusted according to the analysis period for the project (years); the frost-heave rate of the roadbed soil; a "probability of heaving" factor, expressed as a percentage of the project area affected; and the maximum loss of serviceability from Figure 7-3(b). Using these inputs, the loss of serviceability (ΔPSI) attributable to frost action is determined from Figure 7-3(c). This model is the only method known to associate pavement performance in frost conditions with soil and pavement structure characteristics that can be handled at the design stage. It is thus possible to consider improving drainage conditions and reducing the frost penetration and frost susceptibility of subgrade soils to minimize the predicted loss of serviceability for the pavement.

AASHTO now proposes a new mechanistic-empirical pavement design method. Developed as part of a project of the National Cooperative Highway Research Program of the Transportation Research Board, the method is an advanced and a comprehensive design procedure integrated into AASHTOWare, the mechanistic-empirical pavement design software (AASHTO 2015). The mechanistic components of the method include soil and material characterization using resilient modulus for soils and unbound materials and dynamic modulus for asphalt-bound materials. It also includes the computation of pavement structural response to the complete spectrum of axle loads, considering daily, seasonal, and

Month	Roadbed Soil Modulus, M_R (psi)	Relative Damage, u_f
Jan.	20,000	0.01
Feb.	20,000	0.01
Mar.	2,500	1.51
Apr	4,000	0.51
May	4,000	0.51
June	7,000	0.13
July	7,000	0.13
Aug.	7,000	0.13
Sept.	7,000	0.13
Oct.	7,000	0.13
Nov	4,000	0.51
Dec	20,000	0.01
	Summation: Σu_f =	3.72

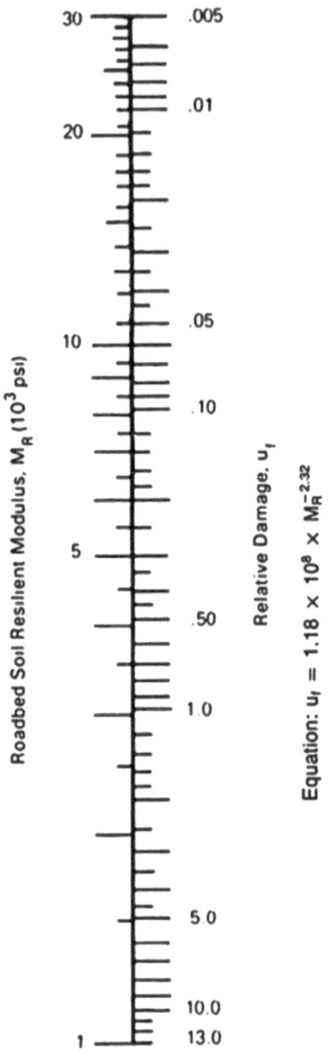

Average: $\bar{u}_f = \frac{\Sigma u_f}{n} = \frac{3.72}{12} = 0.31$

Effective Roadbed Soil Resilient Modulus, M_R (psi) = 5,000 (corresponds to $\bar{u}_f$)

Figure 7-2. Chart for estimating effective roadbed soil resilient modulus for flexible pavements designed using the serviceability criteria.
Source: AASHTO (1993), reproduced with permission.

long-term variations in material properties. The empirical components of the method include distress-specific performance models for fatigue (surface-down longitudinal cracking and bottom-up alligator cracking), permanent deformation

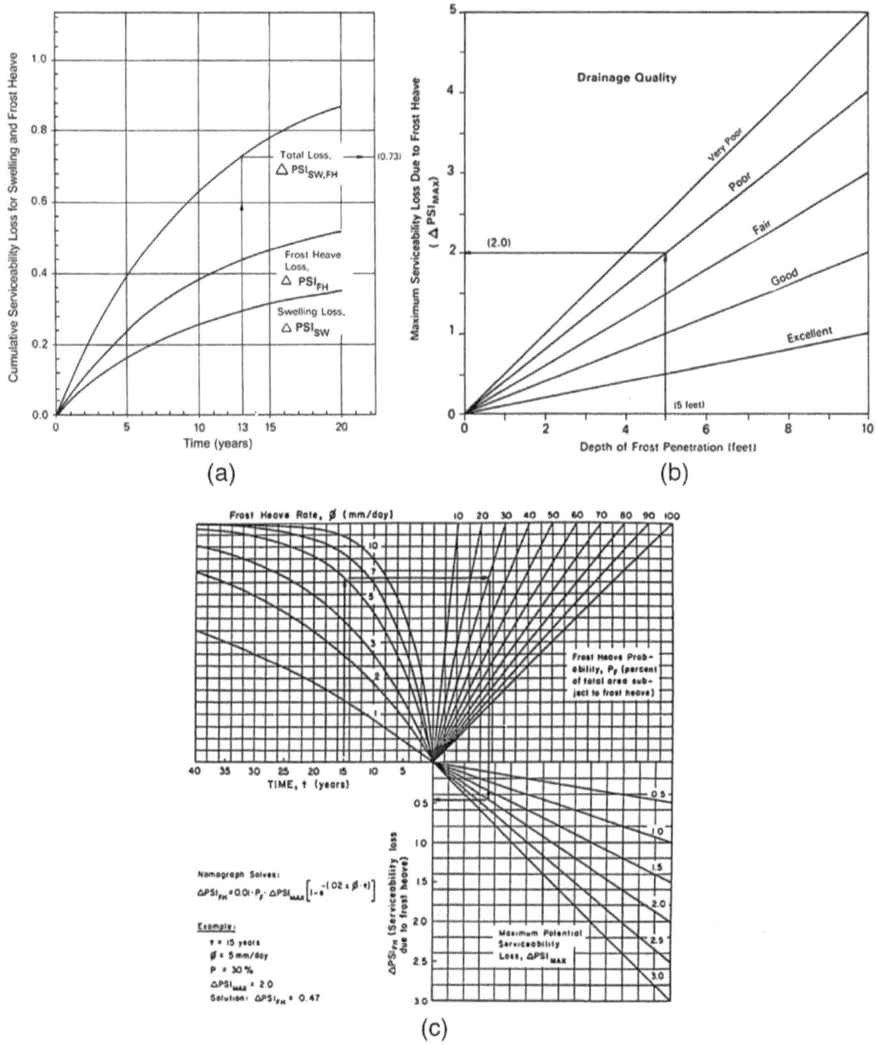

Figure 7-3. AASHTO pavement design procedure for frost protection: (a) concept of serviceability loss due to frost action; (b) maximum serviceability loss as a function of frost penetration and drainage; (c) serviceability loss due to frost action as a function of ΔPSI_{max}, frost-heave probability, and frost-heave rate.
Source: AASHTO (1993), reproduced with permission.

(or rutting) in the granular layers and subgrade soils, thermal cracking, and roughness. Variability of design inputs and performance models is considered in the analysis. The pavement design model uses the Enhanced Integrated Climatic Model to predict temperature and moisture in the pavement structure and subgrade soil (Zapata and Houston 2008, NCHRP 2004). The model is used in damage calculations to assess variation in material properties. It also enables predicting frost depth and frost heave (but frost-heave, per se, is not used in these design procedures). Frost susceptibility is considered to increase heaving in the empirical roughness model used to predict pavement performance. The design method also includes a procedure for calibration to local conditions.

7.2 METHODS USED IN CANADA

The Ontario pavement design method (Ontario Ministry of Transportation 2013, Forsyth 1993, Mackay et al. 1992, He et al. 1996) uses two design procedures based on the principle of partial protection. The first method is presented as a pavement design catalog, where layer thickness varies according to the volume of traffic and the type of soil. The method implicitly proposes protecting against frost action by adjusting the thickness of the granular base and subbase as a function of the volume of traffic, type of soil, and the climatic region considered.

The second procedure, known as Ontario Pavement Analysis of Cost (OPAC), is a rational empirical method based on the principle of serviceability loss as a function of traffic action and environmental factors. This method includes some mechanistic-based procedures to assess the structural adequacy of the pavement structure. It also incorporates a project-level pavement management system to compare different design options. The design steps are illustrated in Figure 7-4 and can be summarized as follows:

1. Deflection at the top of subgrade soil is calculated using the Odemark transform and Boussinesq equations (Doré and Zubeck 2009).
2. Loss of serviceability, ΔRCI, for the analysis period is then calculated using an empirical model relating ΔRCI to subgrade deflection and volume of traffic. (Note: The ride comfort index, or RCI, is based on a scale of 10 points, 10 being a perfect pavement. It can be established subjectively by a group of experts or objectively using a roughness meter.) The model was developed using test data from the AASHTO Road Test in the late 1950s (AASHTO 1993) and from the Brampton road test (Ontario Ministry of Transportation and Communications 1972).
3. ΔRCI for the analysis period, which can be attributed to environmental factors, including frost action, is then calculated using a performance model derived from the Brampton test.
4. Residual serviceability is determined at the end of the analysis period (initial serviceability minus the loss of serviceability). Alternatively, the lifetime expectancy for a structure can be calculated according to a chosen final serviceability value.

The advantage of this approach is to optimize structural design as a function of input parameters such as initial cost, residual value, and user cost. In addition to

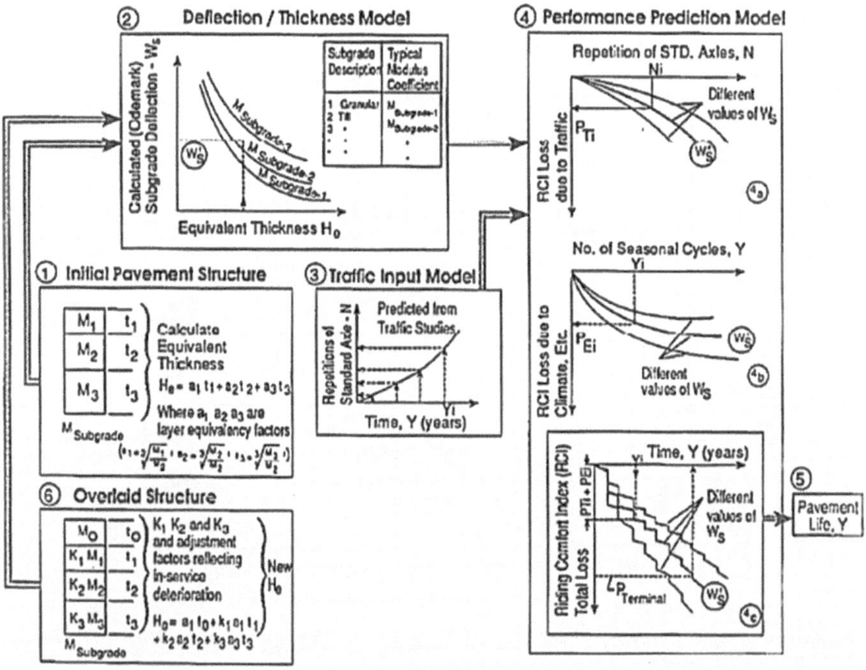

Figure 7-4. Schematic of the Ontario Pavement Analysis of Cost design method used in Ontario.

Source: Ontario Ministry of Transportation, © Queen's Printer for Ontario (2013), reproduced with permission.

using mechanistic principles for the computation of the mechanical response of the pavement system under traffic, the OPAC model uses empirical performance models based on local knowledge and on performance information documented in two important experimental pavement projects. This makes the method highly reliable for pavement design in cold climates.

However, the OPAC model's capacity to consider pavement behavior and the design variables specific to frost action is very limited. The environmental deterioration model relates the loss of serviceability to the deflection of the pavement structure based on estimation of the mechanical response in springtime conditions. This method, therefore, does not allow assessing the impact of mitigation techniques, such as pavement insulation.

The pavement design procedure used by the Québec Ministry of Transportation uses the AASHTO design procedure for structural design. The method was revised in 1995 (Transports Québec 1995) to incorporate the effects of seasonal variations in the mechanical properties of materials by integrating annual damage based on the effective resilient modulus of subgrade soils. Representative monthly resilient modulus values use an empirical function in the 1993 AASHTO method to estimate yearly damage. The 1995 Québec Ministry of Transportation method

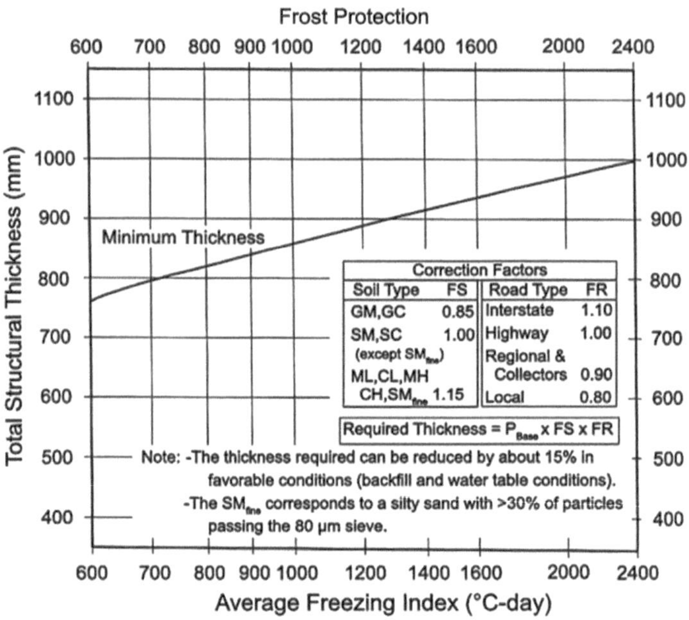

Figure 7-5. Chart used by the Québec Ministry of Transportation for partial frost protection (empirical approach).

Source: Translated from St-Laurent (2006); reproduced with permission of the Ontario Ministry of Transportation.

uses the "Transport Canada frost design chart" (Figure 7-5) to determine the frost protection of the pavement structure. The empirical relationship between the freezing index and the pavement thickness required by Transport Canada for adequate frost protection is based on Canadian experience with frost protection of airfields. The relationship takes into consideration road class and soil type by means of correction factors.

In 2006, the Québec Ministry of Transportation method was further upgraded to include an allowable frost-heave criterion. The method is based on the calculation of frost penetration and frost heave using the segregation potential of pavement materials. The calculated frost heave is then compared to an allowable value based on road class. If the allowable value is exceeded, the designer has to modify the pavement structure by increasing the thickness of the granular layers or by using an insulation layer to reduce frost to an acceptable value (Figure 7-6).

7.3 METHODS USED IN EUROPE

The design method used in Finland (Dysli 1991, Jämsä and Orama 1990) is a mechanistic-empirical approach involving a structural analysis based on the

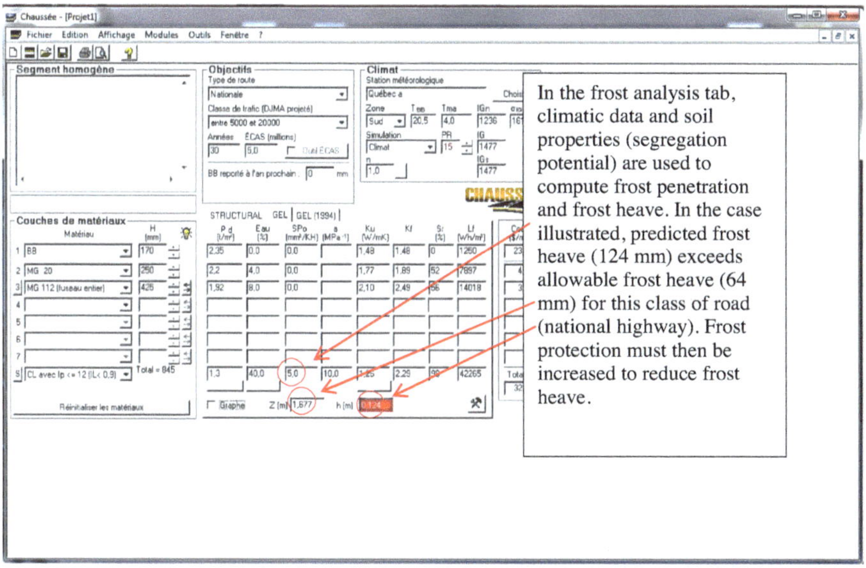

Figure 7-6. Frost analysis method integrated in the pavement design software of the Québec Ministry of Transportation.

Source: Adapted from St-Laurent (2006), reproduced with permission from the Ontario Ministry of Transportation.

Boussinesq and Odemark models, a fatigue damage model, and a procedure to limit frost heave.

The philosophy of the structural procedure is that the pavement must be resistant to critical thawing conditions, which requires specifying the properties of the soil in these conditions. A target bearing capacity is established with a fatigue damage model according to the number of load applications and the service life (typically 20 years). The Odemark model is then used to determine the layer thickness required to obtain the bearing capacity of the pavement according to the stiffness of the soil. The resilient modulus of soil in springtime is used as a design value, which leads to overdesign of the pavement structure.

The minimum thickness to resist frost action is obtained by an empirical relationship [Figure 7-7(top)] established according to the allowable heave [Figure 7-7([bottom)], which is in itself dependent on road class, volume of traffic, type of pavement, frost conditions, and the highest freezing index in 10 years. If the thickness obtained by the structural procedure is less than the thickness determined in the frost procedure, the thickness must be increased or pavement insulation should be considered.

In Norway, frost design is required only for major arterials or roads built in highly frost-sensitive environments (Dysli 1991). In both cases, additional thickness of granular material or insulation is prescribed. The original and interesting aspect of the Norwegian method is the use of a return period to adjust the reliability of the frost design to the importance of the road. The longer the return

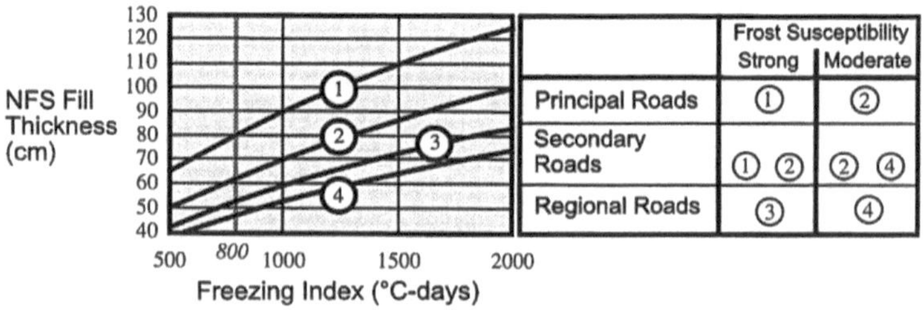

Roadway Classification	Allowable Slope Change [‰]	Allowable Total Frost Heave [mm]
Highways	4-5	40-70
Principal Roadways	5-8	50-100
Regional Roads	7-11	80-200
Secondary Roadways		
Major	9-16	150-300
Minor	15	> 200

Figure 7-7. Frost design method and criteria for the Finnish Road Administration: (top) total thickness required for partial frost protection of pavements in the Finnish pavement design method; (bottom) maximum heave criteria used for the development of the design chart in the top figure.
Source: Translated from Dysli (1991), reproduced with permission from Presses polytechniques et universitaires romandes, Lausanne.

period, the more severe the cold and thus the higher the freezing index. A higher return period will thus be selected for a high-class road to reduce the risk of poor performance. The method uses a frost-protection factor that reflects the risk level in the form of a freezing index over a return period of 2, 5, 10, or 100 years.

The calculation for protection against frost is

$$D = f \times X$$

where

D = Total thickness of the pavement;

X = Frost depth, calculated using the freezing index for a chosen return period; and

f = Protection factor ($0 < f < 1$) determined for different regions in Norway.

For each region, frost-protection solutions are based on an increased thickness of sand and gravel, a granular frost protection layer (coarse-grained crushed rock), thermal insulation, or a “frost accumulation layer” such as wood bark. For insulation, the recommended thicknesses are given for extruded polystyrene, and equivalence factors are provided for other materials. The wood-bark layer can be used in some cases because of its ability to accumulate heat (latent heat contained in absorbed water), which impedes the progression of the frost front.

The French Laboratoire central des ponts et chaussées (now the Institut français des sciences et technologies des transports, de l’aménagement et des réseaux) uses a frost verification method that differs from other international practices (Peyronne and Caroff 1991, Dysli 1991). The design of the pavement structure is established according to traffic (four classes: T3–T0); the bearing capacity of the platform, a layer of engineered soil (three classes: Pf1–Pf3); the mechanical properties of subgrade soil (four classes: S0–S4); the frost susceptibility of subgrade soil (three classes determined by frost-heave tests: SG_n, SG_p, and SG_t); and the air freezing index.

The frost verification procedure uses an allowable freezing index transmitted to the frost-susceptible subgrade soil:

1. Given the pavement structure and the expected severity of the winter, the freezing index transmitted to the subgrade soil is calculated.
2. The freezing index transmitted to the subgrade soil is compared to the allowable freezing index determined from the frost susceptibility class of the soil. The verification is positive if the transmitted freezing index is equal to or less than the allowable freezing index for the soil class.
3. If the allowable freezing index is exceeded, the thickness of pavement granular layers is increased until a satisfactory solution is found.

Figure 7-8(a) illustrates the frost susceptibility classes established by frost-heave tests. Figure 7-8(b) illustrates the French frost verification method, and the table in Figure 7-8(c) specifies the allowable values of transmitted freezing index for different situations.

7.4 DORÉ AND ZUBECK METHOD

Doré and Zubeck (2009) proposed a state-of-the-art method for pavement design in cold climates. It is a five-step mechanistic-empirical procedure adapted to cold environments and is based on two principles:

- The pavement structure must sustain repeated loading across the seasons and over the design life of the pavement without significant loss of structural capacity.
- The pavement structure must survive frost action over the design life of the pavement without significant loss of functional capacity.

Figure 7-9 illustrates the procedure. The following sections summarize the method, and Doré and Zubeck (2009) provide detailed information.

7.4.1 Step 1: Describe Site Conditions

The essential site-specific information includes climatic data and soils and pavement-material characteristics. Mechanical properties of soils and pavement materials should be selected to represent the conditions prevailing during each season. Figure 7-10 illustrates the principle of selecting representative mechanical

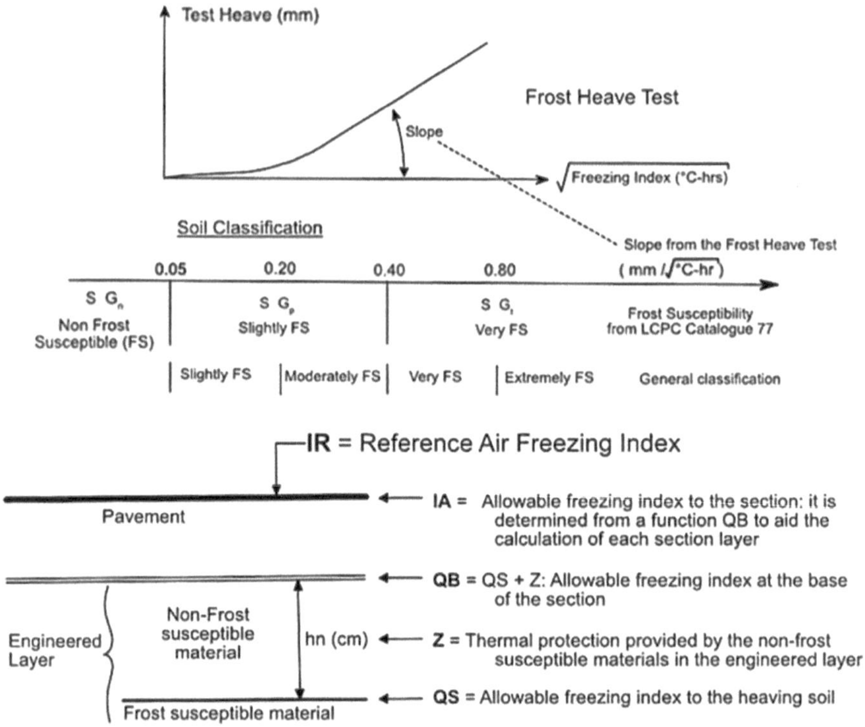

	Engineered Layer Thickness (cm)	0		20		40	
	Fill Type \ Soil Frost Susceptibility	SG_p	SG_t	SG_p	SG_t	SG_p	SG_t
T_1PF_2	GC/GC	185	105	240	150	340	220
	GL/GL	200	110	250	150	350	240
	BC/GC	180	105	225	145	310	215
	GB/GL	130	60	180	100	260	170
	GB/GB	95	40	130	70	200	125

Figure 7-8. Principle of the French frost-protection method: (a) classification of frost susceptibility of subgrade soils; (b) calculation of the transmitted freezing index; and (c) allowable transmitted freezing indices for different frost-susceptibility indices and pavement structure types.
Source: Based on Peyronne and Caroff (1991).

properties for the distinct seasons of the year. The number of seasons and their duration need to be selected carefully to adequately represent the typical variations of material behavior without increasing the burden of pavement design. Seasonal pavement damage analysis generally considers three to five seasons.

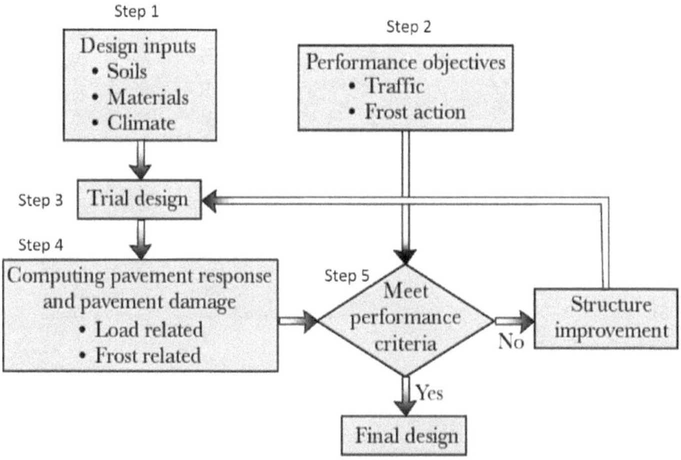

Figure 7-9. Schematic illustration of mechanistic-empirical design procedure adapted for cold climates.
Source: Doré and Zubeck (2009).

7.4.2 Step 2: Establish Performance Objectives

Performance objectives involve selecting a design period during which the structural and functional conditions of the pavement will remain above a selected level of performance or end condition. The selection of performance objectives thus involves two key parameters: the expected design life and the minimum level of service to be provided during this period. The design life can be expressed in terms of years or in terms of number of traffic load applications.

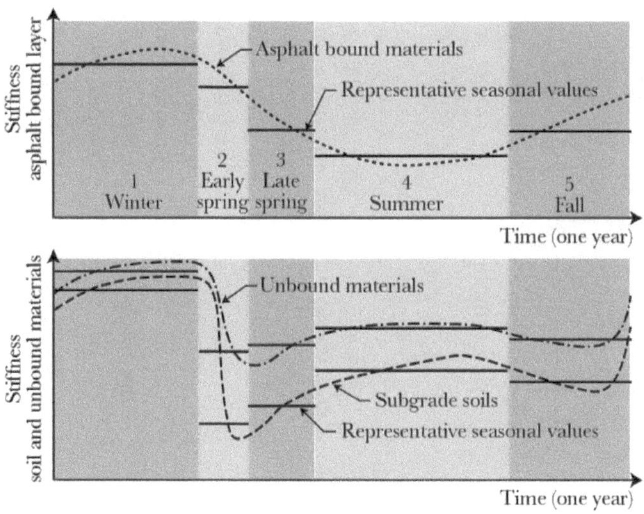

Figure 7-10. Example of selection of mechanical properties representative of seasonal variations for asphalt-bound materials, unbound materials, and subgrade soils.
Source: Doré and Zubeck (2009).

Table 7-1. Performance Indicators for Pavement Design in Cold Regions.

Type of performance	*Performance criteria*	*Related pavement response*	*Recommended terminal condition*	*Comments*
Structural	Fatigue cracking	Tensile strain at the bottom of asphalt-bound layers	20% of wheel path area cracked (Shell and Asphalt institute fatigue cracking models)	Essential information to assess structural performance
	Permanent deformation	Compressive strain (or stress) at the top of the subgrade soil	12 mm (Shell and Asphalt Institute rutting models); 10 mm (TRRL[1], BRRC[2] rutting models)	Essential information to assess structural performance
	Thermal cracking	Can be related to the thermal contraction properties of the asphalt concrete mixture and to the low-temperature tensile strength in a given climatic context	No information available on recommended terminal conditions	Useful information to assess performance of hot mix asphalt concrete (HMAC); primary use is for mix design but can help validate thickness of HMAC layer
Functional	Roughness	Can be related to compressive strain on subgrade soil and variability of soil characteristics (Ullidtz 1998); or to variability of soil frost susceptibility (Doré 1997)	Between 2.0 and 3.5 m/km depending on road classification (based on survey of Canadian agency practices)	Essential information to assess functional performance of pavement; models generally related to frost heave and permanent deformation of pavement structure; few reliable roughness models exist

[1]Transport and Road Research Laboratory (UK).
[2]Belgium Road Research Laboratory.

Performance objectives also involve the selection of structural and functional indicators to represent the minimum acceptable level. In most cases, the terminal service level is included in the empirical models used to predict performance. The performance model should thus be selected to reflect local preferences with respect to the terminal service level. Table 7-1 gives performance indicators for pavements in cold climates.

7.4.3 Step 3: Select a Trial Pavement Structure

Although it is possible to compute the required thicknesses using a mechanistic-empirical design approach, most computer-based procedures will use mechanistic-empirical calculation tools to validate a trial pavement structure and iterate until reaching an optimum. Initial selection of a trial pavement structure is generally based on experience, which generally takes the form of a design catalog.

7.4.4 Step 4: Analyze Pavement Structure

The selected trial pavement structure is analyzed to verify response to traffic and climate loading and to predict related performance. In cold climates, the verification should include

- Analysis of cumulative seasonal fatigue damage,
- Analysis of cumulative seasonal permanent deformation,
- Validation of hot mix asphalt concrete thickness for thermal cracking, and
- Validation of total pavement thickness for allowable frost heave or roughness.

7.4.5 Step 5: Improve Pavement

A pavement structure is considered optimal when the computed damage index over the analysis period is close to but not more than 1 for all damage criteria. Damage index is defined as the calculated damage divided by the allowable level (see guidelines in Table 7-1). Where it exceeds 1 (indicating pavement failure) or is much less than 1 (over-designed), the pavement structure should be modified.

References

AASHTO. 1993. *AASHTO guide for the design of pavement structures.* Washington, DC: AASHTO.

AASHTO. 2015. *Mechanistic-empirical pavement design guide: A manual of practice.* Washington, DC: AASHTO.

Berg, R., and T. Johnson. 1983. *Revised procedure for pavement design under seasonal frost conditions.* Special Rep. No. 83-27. Hanover, NH: USACE, Cold Regions Research and Engineering Laboratory.

Departments of the Army and Air Force. 1985. *Pavement design for seasonal frost conditions: Technical manual TM 5-818-2/AFM 88-6, Ch. 4.* Washington, DC: US Government Printing Office.

Doré, G. 1997. "Deterioration des chausses en conditions de gel: une nouvelle approche previsoinelle." Ph.D. thesis, Départment de Génie Civil, Université Laval.

Doré, G., and H. Zubeck. 2009. *Cold regions pavement engineering.* Reston, VA: ASCE.

Dysli, M. 1991. *Le gel: Et son action sur les sols et les fondations*. Lausanne, Switzerland: Presses polytechniques et universitaires romandes.

Forsyth, R. A. 1993. *Pavement structural design practices: Synthesis of highway practice 189*. Washington, DC: Transportation Research Board, National Academy Press.

Haas, R., W. R. Hudson, and J. Zaniewski. 1994. *Modern pavement management*. Malabar, FL: Krieger.

He, Z., G. Kennepohl, R. Haas, and Y. Cai. 1996. "OPAC 2000: A new pavement design system." In *Proc., Annual Conf. of the Transportation Association of Canada*, Charlottetown, Canada.

Jämsä, H., and R. Orama. 1990. "The measurement of seasonal subgrade properties under existing pavements and the finnish pavement design method." In *Proc., 69th Annual Meeting on TRB Preprint 890619*, Washington, DC.

Ladanyi, B. 1996. *La conception et la réhabilitation des infrastructures de transport en régions nordiques*. Direction des Communications, RTQ-94-07. Montréal: Québec Ministry of Transportation.

Mackay, M. H., D. K. Hein, and J. J. Emery. 1992. "Evaluation of frost action mitigation procedures for highly frost-susceptible soils." *Transp. Res. Rec.* **1362**: 79–89.

NCHRP (National Cooperative Highway Research Program). 2004. *Guide for mechanistic-empirical design of new and rehabilitated pavement structures*. NCHRP Rep. No. I-37A. Washington, DC: Transportation Research Board, National Research Council.

Ontario Ministry of Transportation and Communications. 1972. *Brampton test road: An application of layer analysis to pavement design*. MTC Rep. No. RR182. Toronto, Canada: Ministry of Transportation.

Ontario Ministry of Transportation and Communications. 2013. *Pavement design and rehabilitation manual*, 2nd ed. Toronto, Canada: Ministry of Transportation.

Peyronne, C., and G. Caroff. 1991. *Cours de routes, dimensionnement des chaussées*, 2e éd. Paris: Presses de l'école nationale des ponts et chaussées.

St-Laurent, D. 2006. *Chaussée 2: Logiciel de dimensionnement des chaussées souples. Guide de l'utilisateur*. Montréal: Direction du laboratoire des chaussées, Québec Ministry of Transportation.

Transportation Association of Canada. 1997. *Pavement design and management guide*. Ottawa: Transportation Association of Canada.

Transports Québec. 1995. *Normes: Ouvrages routiers, Tome I: Conception routière Ministère des Transports du Québec*. Montréal: Les publications du Québec.

Ullidtz, P. 1998. *Modelling flexible pavement response and performance*. Lyngby, Denmark: Polyteknisk Forlag.

Zapata, C. E., and W. N. Houston. 2008. *Calibration and validation of the enhanced integrated climatic model for pavement design*. NCHRP Rep. No. 602. Washington, DC: National Cooperative Highway Research Program, Transportation Research Board of the National Academies.

CHAPTER 8
Mitigation of Frost Action

Guy Doré

Two main strategies can be used to protect pavements subjected to frost action (freeze and thaw). The first strategy is to reduce frost heave and thereby control excessive seasonal and long-term roughness development. Limiting frost heave also helps reduce frost–heave cracking and control thaw weakening. The second strategy is to strengthen the pavement structure to make it more resistant to frost–heave forces and thaw weakening.

8.1 FROST–HEAVE MITIGATION

In the case where a pavement structure does not meet the frost protection requirements, it must be improved to ensure the long-term performance of the pavement. There are a wide range of methods available to attenuate the effects of frost action and to improve pavement behavior in cold climates. It is well known that frost heave will happen if three conditions are met: the subgrade soil needs to be exposed to freezing temperatures for a long period; subgrade soils need to be frost susceptible; and water needs to be available near the freezing front to feed the growth of ice lenses. Any action likely to significantly limit one of these factors will improve pavement performance in cold climates. Table 6-1 lists mitigation measures applicable for different design strategies. The following sections provide brief descriptions of some of these methods.

8.1.1 Reduction of Exposure to Freezing Temperatures

Reducing the exposure of subgrade soil to freezing temperatures is the most widely used approach to reduce the effect of frost action. Increasing the thickness of granular layers and adding an insulation layer in the pavement structure can achieve this.

8.1.1.1 Increased Thickness of the Granular Layers

This is by far the most widespread method for reducing or even neutralizing frost heaving, and for reducing the loss of bearing capacity during spring thaw. It consists of adding a nonfrost-susceptible granular subbase layer, or increasing the thickness of the existing layer, between the base layer and the frost-susceptible subgrade soil. The subbase layer is a very important component of the pavement structure in cold and wet environments. It distributes the transmitted load, attenuates differential movements, and generally reduces frost action in frost-susceptible subgrade soils by increasing the depth and therefore the temperature in the subgrade and reducing subgrade water content through improved drainage.

Most existing design methods specify a required total thickness of the pavement structure to resist the effects of freezing. The thickness of bound surface layers and of the base layers is determined from traffic loading requirements. The total thickness of the pavement required is then obtained by adjusting the thickness of the subbase, which is normally made with lower-cost nonfrost-susceptible granular materials.

The remove-and-replace method can be considered as a specific case of this approach when the surface elevation of the pavement is subject to strict limitations. In this case, the subgrade soil is excavated to a depth specified by frost-protection requirements and replaced by an equivalent layer of nonfrost-susceptible granular materials. This approach can also be used when a localized zone of highly frost-susceptible soils is encountered. The pavement structure can thus be adequately protected without raising the final profile of the road. In all cases, the sub-excavated area needs to be adequately drained.

The construction of longitudinal transitions is another example of the remove-and-replace approach. It is widely used to mitigate performance problems related to differential frost heave in areas where the road crosses an abrupt change or boundary between low- and high-frost-susceptibility soil units. Figure 8-1 illustrates a typical transition wedge used in pavement construction.

The main principles used to design a transition wedge are as follows (Dysli 1991):

- Maximum thickness of the wedge should be such that the total thickness of granular protection (*D*) at the contact between the two soil

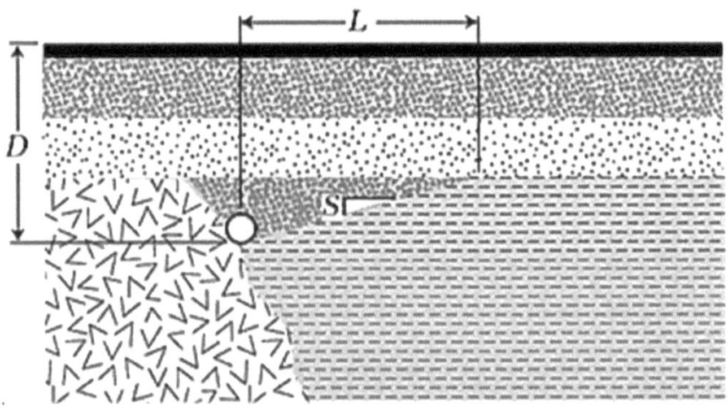

Figure 8-1. Longitudinal transition wedge used to mitigate differential behavior of a road crossing from bedrock (left) to frost-susceptible soil (right).
Source: Doré and Zubeck (2009).

units is equal or greater than the maximum expected frost penetration in granular soils.

- Length of the transition should allow adequate attenuation of differential frost heave between the soil units. The length of the transition (*L*) should take into account traffic speed and expected differences in frost heave between two adjacent soil units. Slopes (*S*) of 20 horizontal to one vertical (20*H*:1*V*) and minimum transition length (*L*) of 20 m are typically recommended for transition wedges.
- Transition wedge must be effectively drained.

Transition wedges are cost-effective when severe differential frost heave is expected between adjacent soil units and when geological contacts between contrasting behavior units are scarce.

8.1.1.2 Pavement Insulation

Pavement insulation is used where widespread severe frost heave and differential frost heave are expected. Contrary to common belief, the insulation layer does not block the progression of the cold front in the pavement system. In fact, it impedes heat flow, reducing heat loss from the pavement system when the pavement surface is exposed to cold air. Thus, soil temperature is likely to remain above freezing throughout the winter. As a result, frost penetration, frost heave, and differential frost heave are minimized, if not eliminated. However, insulation layers are typically made of mechanically weak materials, and the overlying pavement layers need to be designed with special care. Also, a minimum thickness of granular materials is required above the insulation layer to reduce the risk of differential icing at the surface of the pavement. The required thickness depends on specific agency practices.

Different types of insulation materials have been used for road construction. Most pavement insulation materials use air trapped in micropores to reduce thermal conductivity and provide effective insulation. This is the case for expanded and extruded polystyrene and for urethane-based materials. Other materials take advantage of their water absorption capacity to store large quantities of latent heat, which will oppose the progression of the frost front when it reaches the insulation layer. Among other materials, wood residues and peat have been used as pavement insulation materials.

The most frequently used material for pavement insulation is polystyrene boards. With their very low thermal conductivity, relatively high compressive strength, and long-term stability in the pavement environment, polystyrene boards are well adapted for pavement insulation purposes. Extruded polystyrene is the most commonly used and the best performing material. Expanded polystyrene is also used, as a low-cost alternative to extruded polystyrene. But it is more sensitive to water intake, which reduces its insulation properties. This problem needs to be compensated for by increasing the thickness of the polystyrene by a factor of 1.2 (St-Laurent 2006). The major problem with the use of polystyrene boards is the manpower required for their placement. Special care must be given to joints between boards during placement and while covering the boards with granular materials. The use of two board layers allows the joints to be staggered between layers, minimizing the risk of heat loss at joints.

Other products can be used for pavement insulation, including lightweight concrete and lightweight aggregates, which can produce a relatively stiff layer with low thermal conductivity. These materials are also easy to install with standard construction equipment, and it is possible to gradually reduce the thickness of the layer to construct effective transitions. Lightweight aggregates such as expanded clays and foamed glass are becoming interesting insulating materials. But they must be protected against intrusion of fines by a separation geotextile. Also, all of these insulation products tend to be sensitive to moisture intake, and their long-term mechanical and thermal performance has not yet been assessed.

Several types of recovered and recycled materials have also been used for pavement insulation. Laboratory and field tests have used wood residues (bark, sawdust, and wood chips), peat, crumb rubber and chunks, and plastic chips. Most of these materials offer good thermal resistance but typically have poor mechanical properties. They become feasible for pavement insulation when they are available in large quantities and at low cost. They are typically used in the construction or rehabilitation of low-volume local roads, and they require sufficient thickness of granular materials above them to carry the load.

8.1.2 Reduction of the Frost Susceptibility of Subgrade Soils

Another strategy to reduce the effect of frost action in soils is to reduce their frost susceptibility. The methods using this approach include chemical soil stabilization, mechanical soil treatment, and subgrade improvement. In addition to reducing frost heaving or differential frost heaving, these methods are also likely to improve pavement mechanical behavior during spring thaw.

8.1.2.1 Chemical Stabilization

Stabilization by chemical treatment reduces the sensitivity of subgrade soils to frost action. The practical application of chemical treatments is generally limited to the first 300 mm of the subgrade soil. Dysli (1991) describes two techniques frequently used: cement stabilization and lime stabilization.

Cement stabilization binds the particles without modifying their mineralogical properties. As a result, the hydraulic conductivity of the soil is reduced, limiting the possibility of water migrating toward the growing ice lens at the segregation front. It also increases the mechanical resistance of the soil to heave and increases the stiffness during spring thaw. According to Dysli, a gravimetric cement content between 4% and 8% can reduce frost heaving by as much as 90% in the treated layer.

Lime treatment modifies the water absorption properties of the clay particles. It binds the particles and reduces the plasticity of the soil. It takes several months for the treatment to be fully cured and effective. Good penetration of the solute in fine-grained soils makes the technique well suited for soils such as silty clay and clay. It is essential to use freezing tests to determine the optimal dosage of lime for the treatment.

In all cases of chemical stabilization, the materials must have sufficient time to cure and gain strength before they are subjected to freezing temperatures.

8.1.2.2 Mechanical Homogenization

The purpose of this method is to reduce differential frost heave by mechanical homogenization of the top of heterogeneous subgrade soil. Pulvimixers are used to mix to a depth of 900 mm to minimize the risks of differential frost heaving. The technique is used by the state of Minnesota and is documented in their engineering standards (Forsyth 1993). The use of the technique is based on the judgement of the engineer.

8.1.2.3 Improved Subgrade

The improved-subgrade method involves placing or replacing a layer of the existing subgrade soil with borrow materials that have lower frost susceptibility and better mechanical properties. The French pavement design method (Peyronne and Caroff 1991) uses the principle of an engineered soil layer selected and treated to obtain a bearing-capacity level and frost-susceptibility properties specified as a function of road class.

8.1.3 Reduction of Water Availability

Water availability can be reduced by improving the drainage of the pavement structure or by using a capillary barrier underneath a layer of frost-susceptible soil subjected to frost action.

8.1.3.1 Drainage

Effective drainage of the pavement structure can help reduce frost action in two ways. It contributes to minimizing the amount of water present in the pavement and subgrade during freezing by lowering the water table and reducing the availability of water for the growth of ice lenses. But even if good drainage reduces frost heaving, the benefit will be modest. The suction occurring at the segregation front is strong enough to pump water from a water table several meters below the growing ice lens. Effective drainage of the pavement can play an important role when spring thaw occurs, however. Good drainage will aid the rapid dissipation of pore pressure generated by melting segregation ice and will help reduce water content in soil and pavement materials, facilitating rapid recovery of the structural capacity of the pavement.

8.1.3.2 Capillary Barriers

Capillary barriers, which are layers of relatively large-pored material, such as clean coarse sand, clean pea gravel, or a geocomposite of a geonet with geotextile separators to prevent migration of fine soils, reduce upward unsaturated water flow in soils when placed above the water table (Henry and Holtz 2001). Capillary barriers have a low capillary draw and very low unsaturated hydraulic conductivity compared to the soil, thus impeding the upward flow water from reaching soil layers above the capillary barrier. They can also be designed to divert downward, infiltrating water into a drainage system (Henry et al. 2005, Zornberg et al. 2010). Capillary barriers are made of nonfrost-susceptible material. They can be overlain by inexpensive and somewhat frost-susceptible material, as described by Henry et al. (2005). If this approach is used, it is important to ensure that the capillary barrier is placed above the water table at a distance that exceeds the height of capillary rise in the capillary barrier and below the depth of frost penetration required to minimize frost damage.

8.2 STRENGTHENED PAVEMENT STRUCTURE FOR IMPROVED BEARING CAPACITY DURING SPRING THAW

The loss of bearing capacity during spring thaw is, in good part, the result of excess water in pavement granular materials and subgrade soils accumulated by the frost–heave process. Until this water drains from the pavement section and subgrade, bearing capacity will be compromised. Therefore, any action taken to reduce frost heave will also reduce the bearing-capacity loss during spring.

The problem needs to be addressed for the structural design of pavements in cold climates that cannot be fully protected from frost action. Three approaches are usually considered for that purpose. The pavement structure can be designed using soil or material properties representative of spring thaw conditions; an effective resilient modulus for subgrade soil; or a full seasonal damage analysis. In all cases, pavement improvements to resist the loss of bearing capacity during spring thaw involve strengthening the pavement structure by increasing the thickness or the stiffness of the load-bearing layers (asphalt concrete or granular base).

Normal pavement design procedures based on one of these three approaches can assess the thickness and stiffness required to resist spring thaw weakening.

8.2.1 Design Based on Spring Thaw Conditions

This approach involves designing the pavement structure using soil properties corresponding to spring thaw conditions. In most cases, this is done using the soaked California bearing ratio or the saturated resilient modulus as design input for the subgrade soil. Because the design is based on conditions prevailing in subgrade soils for only a few weeks per year, this approach is overly conservative and does not allow the design of optimal pavement structures in cold conditions. It does, however, prevent major damage if pavements are overloaded during that critical period.

8.2.2 Design Based on Effective Subgrade Soil Modulus

The approach proposed by AASHTO (1993) pavement design method is based on an empirical model relating pavement damage to subgrade soil modulus.

Section 7.1 describes this method, which enables optimizing pavement structure based on the variation of subgrade soil properties, including bearing-capacity loss during spring thaw.

8.2.3 Design Based on Seasonal Damage Analysis

This method enables the optimization of pavement structure based on the pavement life consumed during each season of the year. It is generally integrated in most mechanistic-empirical design procedures. Section 7.4 describes the principle of this method.

8.3 CONCLUSIONS

This part of the monograph has summarized the developments made in the area of pavement design in frost conditions over the last few decades. The methods used have evolved from purely empirical approaches to methods based on mechanistic calculations of frost penetration and frost heave. In the context where mechanistic-empirical pavement design is being widely adopted by North American and European transportation agencies, frost design of pavement structures still needs to evolve to become fully mechanistic-empirical. Frost heave is the response of pavements to frost action, and it can be calculated using mechanistic methods based on the segregation potential or other equivalent theories. The remaining challenge is to link the mechanistic response to performance by developing frost-damage functions using frost heave as dependent variable. These functions would enable the determination of allowable frost heave for any given site conditions, making optimization of pavement structures possible. This requires that frost action, including frost heave and spring thaw weakening, be adequately addressed in design procedures.

References

AASHTO. 1993. *AASHTO guide for the design of pavement structures.* Washington, DC: AASHTO.

Doré, G., and H. Zubeck. 2009. *Cold regions pavement engineering.* Reston, VA: ASCE.

Dysli, M. 1991. *Le gel: et son action sur les sols et les fondations.* Lausanne, Switzerland: Presses polytechniques et universitaires romandes.

Forsyth, R. A. 1993. *Pavement structural design practices: Synthesis of highway practice 189.* Washington, DC: Transportation Research Board, National Academy Press.

Henry, K. S., and R. D. Holtz. 2001. "Geocomposite capillary barriers to reduce frost heave in soil." *Can. Geotech. J.* **38** (4): 678–694.

Henry, K. S., J. P. Olson, S. Farrington, and J. Lens. 2005. *Improved performance of unpaved roads during spring thaw.* ERDC/CRREL TR-05-01. Hanover, NH: USACE, Engineer Research and Development Center.

Peyronne, C., and G. Caroff. 1991. *Cours de routes, dimensionnement des chaussées,* 2e éd. Paris: Presses de l'école nationale des ponts et chaussées.

St-Laurent, D. 2006. *Chaussée 2: Logiciel de dimensionnement des chaussées souples. Guide de l'utilisateur.* Montréal: Direction du laboratoire des chaussées, Québec Ministry of Transportation.

Zornberg, J., A. Bouazza, and J. S. McCartney. 2010. "Geosynthetic capillary barriers: Current state of knowledge." *Geosynthetics Int.* **17** (5): 273–300.

PART III

Case Studies

Introduction

Jeremiah Drage

Examples of frost heave in action are often found in nature and in the constructed world in cold regions. Residents of cold regions sometimes need only look out their window to see examples, such as uneven fence posts caused by vertical jacking. Others may see examples when they drive on the highway and experience a bump between a fill embankment and a bridge structure. Along the greenbelt trail near my house are shallow culverts that have turned into "speed bumps" over the years. This case study section presents more examples, and the previous sections of the monograph discussed the processes behind these examples.

Frost action on structures is often the result of an embedment depth that lacks adequate resistance capacity relative to the applied frost load, or of inadequate or nonexistent measures to mitigate frost penetration. For structures that do not require significant load capacity to meet their intended purpose, such as fence posts, equipment protection bollards, road signage, or culverts, the owner may accept some frost jacking rather than investing in the cost to design and build the structure to mitigate frost effects. Other times, an important structure may be damaged by frost action. A condominium development near Silverthorne, Colorado, was significantly damaged by frost action on a structural slab foundation. This example is presented as Case Study 1, with excerpts from a paper by Overton et al. (2009).

Frost action on road sections can have little noticeable effect if it occurs in a uniform manner, as discussed earlier in this monograph. However, when frost action causes differential movement along the alignment, drivers may be disturbed by an uncomfortable bump; or worse, the undulating surface may compromise traveling safety. Case Study 2 presents methods commonly used by the Québec Ministry of Transportation to analyze the performance of existing road alignments and provides an example of the analysis method (St-Laurent and Brillant 2016).

Case Study 3 further investigates of the mechanisms of frost heave on pavement sections with a look at the airfield pavements at Fort Greely, Alaska. This work includes ground penetrating radar surveys, boreholes, and soil moisture and grain size distribution analysis of five sites at Allen Army Airfield.

References

Overton, D. D., J. D. Nelson, and K. Chao. 2009. "Analyses of frost-mitigation under post-tensioned slabs." In *Proc., Cold Regions Engineering Conf.*, Duluth, MN, 119–129.

St-Laurent, D., and G. Brillant. 2016. *Study of the frost susceptibility of existing pavement.* Montreal: Québec Ministry of Transportation.

CASE STUDY 1

Analyses of Frost Migration under Building Perimeter Slabs

Daniel D. Overton
John D. Nelson
Kuo-Chieh Chao

INTRODUCTION

Twenty-one wood-frame condominium buildings were founded on post-tensioned slabs-on-grade at a site in Silverthorne, Colorado. The site is at an elevation of 2,562 m (8,700 ft) in the Rocky Mountains and borders the Blue River. The preconstruction soils report recommended that a frost-protection beam be constructed around the perimeter of the building to a minimum depth of 1,220 mm (48 in.). Evidence of frost heave was recorded in the first winter after construction.

We conducted a forensic geotechnical investigation to identify soil conditions at the site, determine the causes of distress to the structures, and determine a remediation plan.

PROJECT DESCRIPTION

The soils at the site consist of silty gravels, silty sands, sandy silt, and silty clay. Because the distress was caused primarily by frost heave, characterization of the soils concentrated on the soils in the top 2.1 m (7 ft) below the ground surface. Twenty test pits were excavated to depths of 1.5 to 1.8 m (5 to 6 ft) across the site. Figure C1-1 shows the section and subsoil profile analyzed for this study.

Post-tensioned slabs were used as foundations and were designed with a 1,220 mm (48 in.) deep frost beam around the heated portions of the structure.

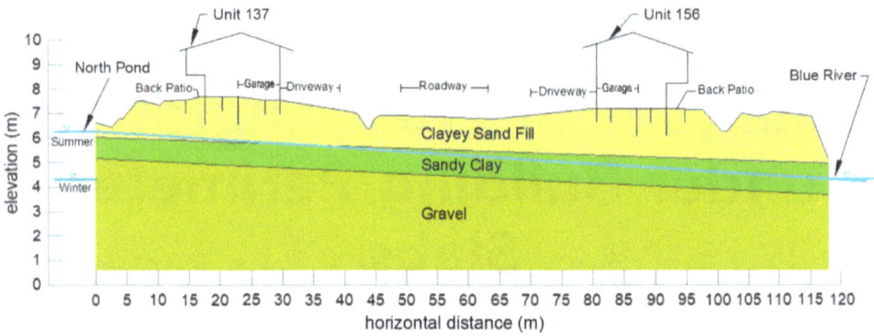

Figure C1-1. Section and subsoil profile analyzed in the study.
Source: Overton et al. (2009).

However, these buildings were designed with an unheated garage and an open rear patio, and the floor of the patio was cast as an integral part of the post-tensioned foundation. The unheated rear patio supported a column for the second story. Although design frost depth was 1,070 mm (42 in.) at the site, the frost beams for the unheated garage and patio sections were designed as only 660 mm (26 in.) deep.

During the first winter after the buildings were constructed, the owners observed heave of the post-tensioned foundation system. This movement was most pronounced in the unheated back patio slab, with a lesser amount of heave in the garage slab. The maximum heave was approximately 76 mm (3 in.) during the middle of the winter and decreased during the following spring. The heave of the rear patio slab was transferred to the second story through the column supported by the unheated patio slab. Some heave of the patio was also transferred through the slab to the heated portions of the ground floor.

FROST MODELING

The depth and distribution of freezing temperatures in the soil were analyzed using the finite element computer program VADOSE/W, Version 6.13 (GEO-SLOPE 2004). This program can be used to model water migration and variation of temperature with time in soil and bedrock in both saturated and unsaturated conditions. It is possible to consider infiltration from precipitation or surface-water ponding, evaporation, and heat flow.

The section and subsurface profile extends through the garage area and the unheated patio slab. The hydraulic boundaries of the section shown in Figure C1-1 are the Blue River on the right-hand side and irrigation ponds on the left-hand side. The ponds are drained in the winter, causing the groundwater level to vary seasonally. A constant-temperature boundary of 10 °C (50 °F) was specified at a depth of 7.0 to 7.6 m (23 to 25 ft). This corresponds to the soil and groundwater temperature at that depth.

Climate data were obtained from the National Oceanic and Atmospheric Administration's Station Dillon 1E. Daily minimum and maximum temperatures and precipitation values were the model input parameters. The model was calibrated using climate data from September 1, 2002, through April 30, 2003. This period includes the time when we performed the forensic geotechnical field investigation, allowing the temperature data measured in the test pits to be used to calibrate the model.

After the model was calibrated, it was used to aid in designing remedial measures. For design purposes, climate data for the winter of 1978–1979 were used. That winter was the coldest winter since 1970, with an air freezing index of 1,267 °C-days (2,312 °F-days). The air freezing index for the 25-, 50-, and 100-year return periods is 1,248, 1,280, and 1,308 °C-days (2,279, 2,337, and 2,387 °F-days), respectively. Linear interpolation between these values indicates that 1978–1979 corresponds to a return period of about 39 years.

For unpaved areas, surface boundary conditions were modeled as fluctuating daily temperatures. The building floors were considered to be constant-temperature, no-flow boundaries. The temperature of the slab beneath the living areas of structures was measured to be 8.3 °C (47 °F) at the edge and 10.6 °C (51 °F) at the interior [more than 610 mm (24 in.)] from the edge of the slab). The garage floor temperatures were fairly constant at 1.7 °C (35 °F).

The soils modeled in VADOSE/W were chosen as representative of the site based on our test pit excavations and review of documents pertaining to this site. Table C1-1 presents the soil properties used in the model. Overton et al. (2009) fully describe the model calibration and validation.

RESULTS

Temperatures under the garage, living area, and back porch were modeled for September 1, 2002, to April 30, 2003, the period when the distress occurred. The model results indicated that frost would develop beneath the back patio slab to a depth of 1.0 m (3.2 ft). That profile was then modeled for September 1978 through April 1979, which was the coldest winter in the Silverthorne area since 1970. Figure C1-2 presents the model results: frost would develop to a depth of 1.1 m (3.7 ft) beneath the back patio slab.

Frost heave was predicted using the frost-heave model PCHeave (Sheng 2003) and input parameters from the VADOSE/W analyses. PCHeave is a one-dimensional model that simulates frost heave and ice lensing. The section and subsurface profile shown in Figure C1-1 were used to construct a one-dimensional soil column with a depth of 2.6 m (8.4 ft) at the back patio. Modeling was performed using PCHeave and the climate data for the 2002–2003 and 1978–1979 winters, as previously discussed.

Table C1-1. Soil Parameters Used in the Frost Analyses.

Soil type	*Saturated hydraulic conductivity (cm/s)*	*Permeability anisotropy ratio* (K_h/K_v)	*Saturated vol. water content (vol./vol.)*	*Residual vol. water content (vol./vol.)*	*Soil mineral thermal conductivity (J/s m °C)*	*Soil mineral specific heat capacity (J/g °C)*	*Bulk Density* (Mg/m^3)
Clayey sand fill	1.01×10^{-4}	1	0.45	0.16	0.52	0.84	2.01
Sandy/silty clay	1.01×10^{-4}	1	0.29	0.18	0.87	0.84	1.85
Sandy clay	1.01×10^{-4}	1	0.29	0.18	0.87	0.84	1.90
Gravel	1.1×10^{-2}	1	0.30	0.05	8.65	1.13	2.08
Polystyrene	9.3×10^{-11}	1	0.1	0.1	0.04	1.13	0.03
Asphalt	9.3×10^{-11}	1	0.1	0.1	1.49	1.67	2.32
Concrete	9.3×10^{-11}	1	0.1	0.1	1.73	0.88	2.40

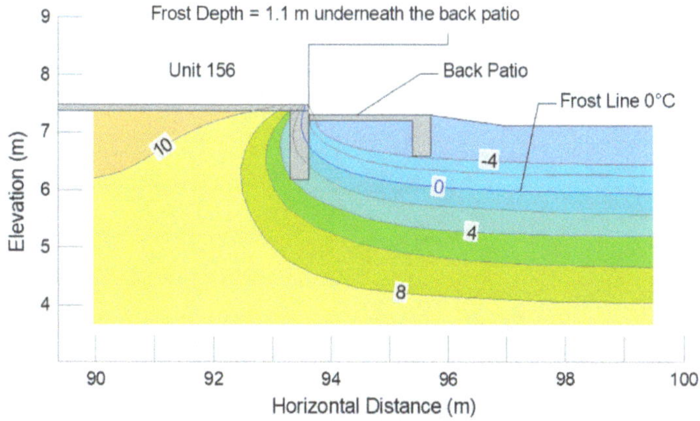

Figure C1-2. Subsoil temperature profiles for the constructed foundation with 1978–1979 climate data.
Source: Overton et al. (2009).

The surface-temperature data used in the PCHeave analyses were taken from the results of the VADOSE/W modeling. A constant-temperature boundary was specified at a depth of 7.0 to 7.6 m (23 to 25 ft). The soil properties listed in Table C1-1 were used in the PCHeave models. An additional parameter, the ratio of the unfrozen water content to the initial water content, was estimated to be 20% at a temperature of −1 °C and was based on the soil type and information provided by Sheng et al. (1995a, 1995b). The clayey sand fill and sandy clay were specified as frost-susceptible materials.

The analysis predicted a maximum heave of 152 mm (6.0 in.) beneath the back patio in 2002–2003 and 173 mm (6.8 in.) in 1978–1979. Observation of three to six inches of heave in the asphalt areas during the early spring of the following year confirmed the magnitude of heave that was predicted. This further supports the credibility of the method.

REMEDIATION

Various remediation schemes to provide adequate insulation beneath the back patio slabs were considered. The design year that was used was the winter of 1978–1979, as discussed previously. Various remediation schemes were considered and modeled using VADOSE/W. Figure C1-3 shows the final insulation plan we proposed for the back patio.

Our analyses determined that a 1,220 mm (48 in.) deep insulated sister beam and additional insulation were needed to eliminate frost penetration beneath the back patio slab. Icynene insulation 305 mm (12 in.) thick was proposed under and along the inside of the existing frost beam. In addition, 51 mm (2 in.) of insulation

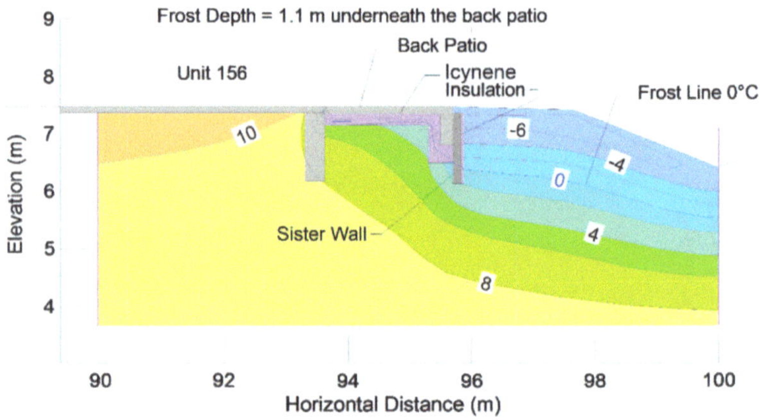

Figure C1-3. Subsoil temperature profiles for the proposed plan with 1,220 mm deep sister wall and additional insulation, 1978–1979 climate data.
Source: Overton et al. (2009).

was required on the outside of the sister wall, and 203 mm (8 in.) of insulation under the patio slab. The model results show that the frost is confined to the insulation and that the soils beneath the insulation do not freeze.

CONCLUSIONS

The original design of the post-tensioned slab failed to consider the effects of frost under the unheated portions of the slab. This failure was evident in the design by the reduction of the frost-beam depth from 1,220 mm (48 in.) in the heated area to only 660 mm (26 in.) in the unheated area. In reality, the depth of the frost beam and efforts to minimize frost in the unheated areas warranted more effort than in the heated areas.

It is not unusual for exterior patios to be cast as integral parts of post-tensioned slab foundations. In many cases that slab also provides support for an overhanging portion of the structure. This does not pose a problem if the soil under that portion of the slab is not susceptible to freezing or frost heave. However, when freezing of the subsoil is possible, it is necessary to insulate both the underside of the slab and the exterior frost beam. If an unheated area is to be located adjacent to a heated area and if frost is possible, we recommend that the unheated slab be cast separately. It should be isolated from the adjacent structure, and a separate nonfrost-susceptible foundation should be used to support any overhanging structures. Alternatively, the soil beneath the unheated slab can be excavated and replaced with nonfrost-susceptible material.

The VADOSE/W analyses that this study presented provided very valuable information, particularly in being able to evaluate various insulation schemes.

The ability to calibrate and validate the models for an actual situation demonstrated the credibility of this method of analysis.

References

GEO-SLOPE International. 2004. *GEO-STUDIO VADOSE/W software package for seepage analysis, version 6.13*. Calgary, Canada: GEO-SLOPE International.

Overton, D. D., J. D. Nelson, and K. C. Chao. 2009. "Analyses of frost-migration under post-tensioned slabs." In *Proc., 14th Conf. on Cold Regions Engineering*, Duluth, MN.

Sheng, D., K. Axelsson, and S. Knutsson. 1995a. "Frost heave due to ice lens formation in freezing soils. 1: Theory and verification." *Nord. Hydrol.* **26**: 125–146.

Sheng, D., K. Axelsson, and S. Knutsson. 1995b. "Frost heave due to ice lens formation in freezing soils. 2: Field application." *Nord. Hydrol.* **26**: 147–168.

Sheng, D. K. 2003. *PCHeave model, version 4.0*. Callaghan, Australia: Univ. of Newcastle, School of Engineering.

CASE STUDY 2

Frost Susceptibility of Existing Pavement

Denis St-Laurent
Gilles Brillant

INTRODUCTION

Most road maintenance and improvement techniques have only mild success on pavement that is deformed year after year by frost. This must be kept in mind when designing pavement structures or choosing repair options for them. The selection of maintenance options is easier for existing pavement, as the pavement's behavior in winter can be observed and some additional basic data can be collected.

Frost causes more or less uniform heaving (Figure C2-1). Such deformation is seen toward the end of winter, when frost reaches its maximum depth, and is caused by the formation of ice lenses, which are themselves caused by cryosuction in frost-susceptible soil.

Frost heave can be reproduced in the laboratory and can be quantified using certain concepts, such as segregation potential (Konrad and Morgenstern 1983). There are also some theoretical methods to calculate it according to climate and the layers of material. The Saarelainen (1992) model is undoubtedly one of the simplest. The Québec Ministry of Transportation relies on these models to analyze the effects of frost, in particular with the help of the software program Chaussée 2 (St-Laurent 2006). The following is a summary of the frost analysis tool's key features, a step-by-step method for engineering studies on existing frost-susceptible pavement, and a real-life case study.

CALCULATION METHODS

The first version of the frost analysis tool was developed in 2000, and its performance was then validated using in situ measurements. The tool was made available in the second version of Chaussée, which has been available on the

(a) (b)

Figure C2-1. Frost heave: (a) summer, (b) winter.

Québec Ministry of Transportation website since 2006. This software also includes a structural calculation module based on AASHTO's (1993) pavement design guide.

The frost analysis module is mainly a local implementation of the approach originally developed in Finland by Saarelainen. This model determines frost depth based on the heat balance at the freezing front. This approach allows the progression of the freezing front to be simulated for every day of winter. Frost heave is determined using the segregation potential (SP).

The analysis takes into account the daily temperatures of a given winter. A sinusoidal pattern of daily temperatures can also be derived from freezing indexes. The analysis also uses the average yearly air temperature. Users must choose the nearest weather station and can choose either the desired return period for the targeted winter or the daily temperatures of a particular winter.

Pavement structures are modelled using the following characteristics for each layer: thickness, dry density, mineralogy of particles, water content, asphalt content, specific surface, SP at zero overburden, and a constant related to the soil's frost-heave sensitivity to overburden pressure. Some of these data are used to first analyze the air and ice content, thermal conductivity, and latent heat of fusion of each layer. Ice content is determined using Anderson and Tice's model (1972), and thermal conductivity is calculated with the Côté and Konrad model (2005). Thermal conductivity may also be calculated using Kersten's model (1949). The software comes with a library of default values for all the properties of various material types.

The analysis gives an instant prediction of frost depth and frost heave. Users can adjust pavement structures to keep frost heave below an acceptable threshold. All parameters can be modified easily, and sensitivity studies can be performed very quickly. The program is flexible enough to accommodate analyses of all characteristics encountered in practice. The most important parameters are immediately visible, with the rest shown in subordinate windows.

The software's default SP values are safe and realistic for most conditions encountered in the province of Québec in Canada. However, the SP value is one of the most influential parameters; it is therefore recommended to take the time to establish this value accurately for the best results. This can be done directly in the laboratory, or indirectly using a characterization of the fine particles: sedimentation analysis, specific surface, and Atterberg limits (Konrad 2005). The SP of an existing pavement can also be determined by using inverse calculations based on in situ heave measurements—in practice, this approach should be favored where conditions allow. The next section provides a detailed methodology for analyzing existing pavement. A real-life example follows.

METHODOLOGY FOR EXISTING PAVEMENT

Following are the steps of the preferred analysis method. Any equivalent tool can replace Chaussée 2.

1. Do a visual survey and mark the locations to be measured. An assessment in winter (February–March) is preferable to better determine the areas sensitive to frost.
2. Collect stratigraphic data. In general, a few soil samples are taken from boreholes, and several analyses (grain size, water content, and Atterberg limits) are performed. Such data may be available from previous studies.
   - Optional: Insert a frost tube in one of the holes to measure the frost depth and make complementary soil analyses (sedimentation and methylene blue value) or laboratory SP measurements.
3. Measure surface elevations in late winter (February–March) and in summer (May–November). A land survey with a rod and level is the most common method (Figure C2-2). Check the survey's results for precision by going back to the starting point. An error of closure of less than 3 mm is common;

(a) (b)

Figure C2-2. Measuring frost heave, (a) using a rod and level; (b) using a portable profilometer.

repeat the survey if the error is over 10 mm. Winter and summer measurements must be taken at exactly the same location, which requires a marker of some sort (paint or surveyor's nails). It is important that elevations be measured against a fixed point that is unaffected by frost (rock, building, or other structure) to calculate the difference between the winter and summer values. When no fixed point is available at the site, a metal rod with an independent flexible casing is sometimes used, anchored to the bottom of a hole over 3 m deep using a disk welded to the rod's base.

- Optional: In addition, use profilometer surveys for a higher number of points (continuous measure), and calculate the difference in ride quality, $\Delta IRI = IRI_{winter} - IRI_{summer}$ (IRI: international roughness index). Preferably, use a profilometer capable of measuring the true, unfiltered profile, which is the case with the portable devices equipped with an inclinometer.

4. Plot the results on a graph. Frost heave is calculated as the difference between winter and summer elevations. Identify homogeneous areas, interpret the results, draw conclusions, and determine the paths to take. Choose characteristic points for the design calculations.
5. Import the climate data relative to the winter encountered in Step 3. The Environment Canada website publishes daily temperatures monthly. The Chaussée 2 user guide explains the procedure and file format for importing these temperatures.
6. Calibrate Chaussée 2 to ensure that the results are representative of the site studied. Do this by choosing the winter during which the measurements were taken and entering the other available data (material types, thicknesses, water contents, and dry densities). The program gives the saturation percentage to help choose a dry density that is compatible with the water content used. The soil strata can also be subdivided according to their makeup or the water table depth. Adjust the soil's SP as needed to ensure a calculation representative of the behavior observed at the site (frost depth and heave).
7. Record the soil characteristics obtained in the previous step. Choose the targeted or design winter and simulate the layered system corresponding to the possible pavement repair options. Adjust these options' variables so that frost heave is within the acceptable limit—50 to 70 mm, according to the current standards of the Québec Ministry of Transportation.

APPLICATION OF THE METHOD

The pavement section is on Route 283 in Saint-Paul-de-Montminy, in the Appalachians. This section of pavement was reconstructed in 2000 after it was damaged by frost (Figure C2-3). It has a north-facing 7% slope. The figure shows some of the section's main cracks. Visual markers (pairs of surveyor's nails) at

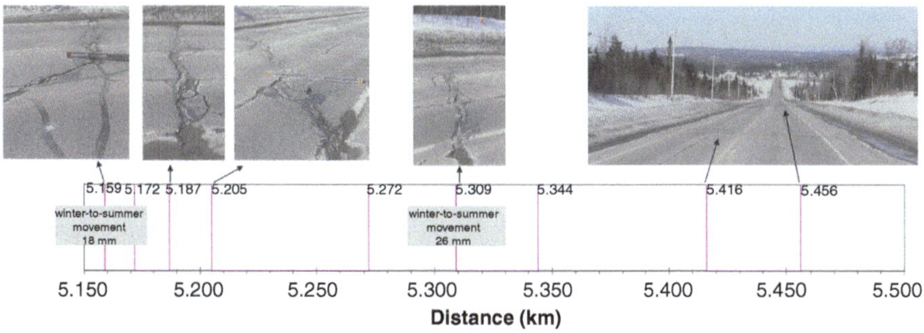

Figure C2-3. Main transverse cracks.

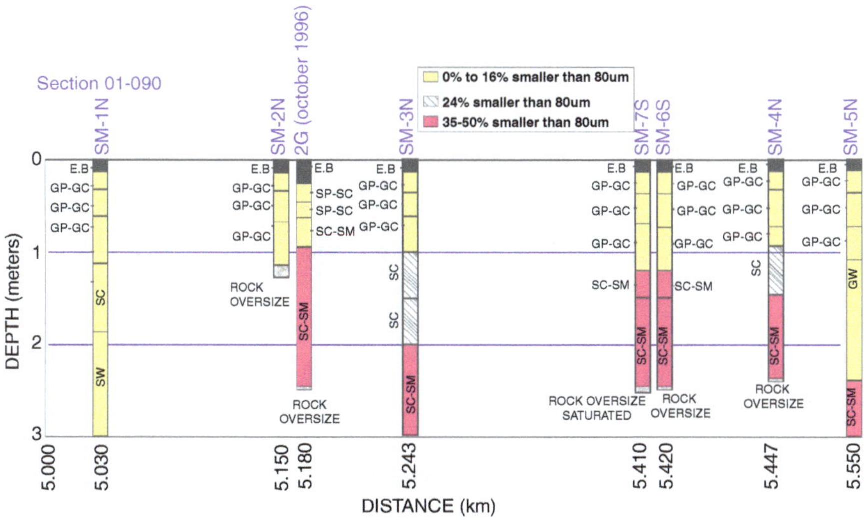

Figure C2-4. Results of the boreholes.

some cracks showed that cracks were reacting to frost. The cracks were 15 to 26 mm wider in winter than in summer. No crack sealant can prevent that movement. Figure C2-4 shows the borehole results. Basement rock is at less than 2.5 m in several places, and the potentially frost-susceptible supporting soil is made up of 35% to 50% fine particles smaller than 80 μm (samples classified as silty clayey sand, SC-SM, liquid limit 21, plasticity index 5). The materials sampled are shown according to the range of fines indicated in the legend.

Figure C2-5 shows the profilometry measurements, including the longitudinal profiles in summer and winter, measured in the northbound lane. The figure also shows the difference between the two measurements, which represents heave. The profiles were reoriented along the horizontal: the last point was adjusted to the same elevation as the first to reduce the scale of the elevations and to better see the deformations. The dots represent measurements taken with the rod and

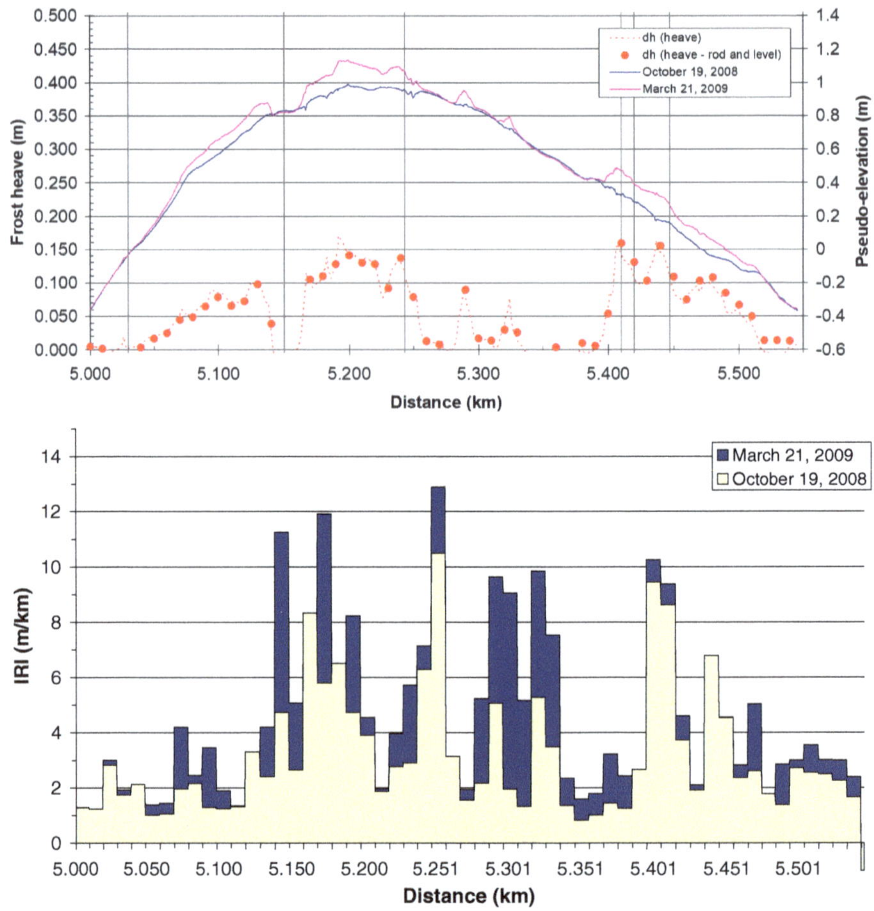

Figure C2-5. Profilometry results: frost heave (top) and international roughness index (bottom).

level method, and the curves represent measurements taken with a portable profilometer. The vertical lines show the location of the boreholes whose stratigraphy is known. IRI values, calculated every 10 m, are also illustrated to show variations in ride quality. Measurements from the other lane are not given; they were very similar to those shown.

Many deductions can be made even before using Chaussée 2. Frost heave varies widely (0 to 170 mm) over the length of the studied area. Deformations seem for the most part related to poor soil–rock transitions. The borehole at distance 5.150 reached rock at a depth of 1.2 m; heave was non-existent there—which is logical—and we can assume that the 20 m section with zero heave was delimited exactly by that rock outcrop. The soil–rock transition there is considered deficient, with IRI of 4.7 m/km and 11.3 m/km in October and March, respectively. IRI is poor at many other points where the heave and profiles are clearly irregular (e.g., distance 5.250, 5.300, and 5.400). Overall, frost heave at its worst was up to 170 mm.

Cracks and profile distortions will surely reappear next winter if nothing more than surface recycling or asphalt paving is done. Eliminating the problem requires more serious repair, such as further excavation or the insertion of polystyrene insulation panels. Figure C2-5 shows locations unaffected by frost heave that can serve as the starting and end points for panel placement without any risk of differential heave. The boundaries of the area being studied (5.050 to 5.520) may represent very good starting and ending points as they correspond to frost heave already vanishing. This avoids any potential differential frost heave issues. Alternatively, polystyrene panels can be started or ended between different frost-susceptible areas if the budget allows only one or some portion of the section to be repaired; the absence of heave between frost-susceptible subsections would ensure transition points without differential heave. The absence of frost heave at the boundaries of the site studied is explained by the thickness (2.5 m and over) of the high-quality material (less than 15% fine particles) revealed in the samples taken at distances 5.020 and 5.550.

Chaussée 2 is used to complete the analysis of the situation. For the case in question, we chose the initial stratigraphic profile from borehole SM-7S, at 5.410, where on March 19, 2009, we measured a frost heave (h) of 139 mm. A frost tube installed nearby (km 5.243) indicated a frost depth (Z) of 2.37 m. Figure C2-6

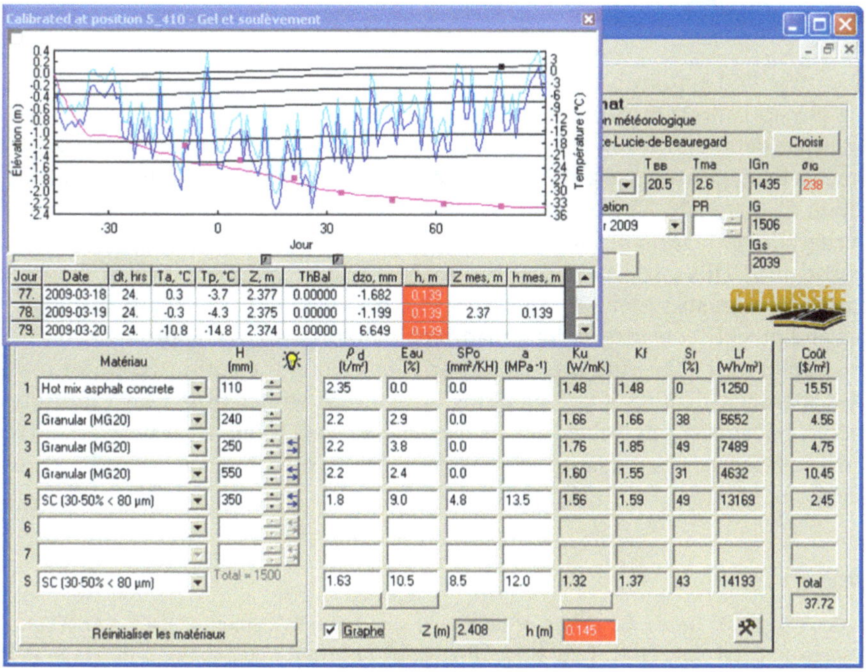

Figure C2-6. Chaussée 2 simulation, calibrated for the borehole at 5.410.

shows the first calculation. The figure shows the layers' data (material, thickness, and water content) and the SP at zero overburden (SP_0 of 4.8 and 8.5) found by trial and error, so that the analysis took into account both the progression of the freezing front and the frost heave measured on the road. The plot shows the simulated temperatures, freezing front, and frost heave in addition to the frost depth measured on seven successive days and the heave seen on March 19. The fit between the simulated (solid lines) and measured (dots) frost heave values is agreeable. This gives us a representative model of how the pavement at this location behaves. The only action left is to choose the targeted design winter, for instance a winter with a return period of 12 years, which is colder than 2009's winter, and modify the pavement structure according to the possible pavement repair options. The program will output results in terms of expected frost heave. In this case, the study suggests that the pavement should be raised by nearly 1 m (an option we rejected), the area beneath the pavement to a depth of 2.25 m should be reconstructed, or the area beneath the pavement to a depth of 650 mm should be reconstructed to insert 50 mm thick extruded polystyrene panels. The minimal depth for the positioning of the polystyrene is determined by the control of icing on the surface (Côté and Konrad 2006). The three options are equivalent in terms of expected performance, but the last is the most cost-effective and the easiest to implement. The material already on-site can also be recycled.

CONCLUSIONS

The presented approach to choosing a road repair option is based on modelling the physical mechanisms at play in freezing weather. It can be used with easily available data, and leads to quantifiable performance indicators. In terms of a practical engineering exercise, slight inaccuracy may be present in the characterization of materials. The different sources of error tend to cancel each other out during the calibration of the model with in situ measurements of frost heave and frost depth. This guarantees that the analysis will be representative of the soil–pavement–climate system. The designer thus has everything needed to analyze the site and quickly compare various road-repair alternatives. Using profilometry has the advantage of showing the spatial variations along the entire section of road. The approach requires some planning, as it needs winter frost heave measurements, but it is highly valued for its usefulness in analyzing frost problems. The Québec Ministère des Transports has successfully used this approach on many occasions.

References

AASHTO. 1993. *Guide for design of pavement structures.* Washington, DC: AASHTO.

Anderson, D. M., and A. R. Tice. 1972. "Predicting unfrozen water contents in frozen soils from surface area measurements." *Highway Res. Rec.* **393**: 12–18.

Côté, J., and J.-M. Konrad. 2005. "A generalized thermal conductivity model for soils and construction materials." *Can. Geotech. J.* **42** (2): 443–458.

Côté, J., and J.-M. Konrad. 2006. "Granular protection design to minimize differential icing on insulated pavements." *Can. Geotech. J.* **43** (3): 260–272.

Kersten, M. S. 1949. *Laboratory research for the determination of the thermal properties of soils.* Technical Rep. No. 23. Minneapolis: Univ. of Minnesota Research Laboratory Investigations, Engineering Experiment Station.

Konrad, J.-M. 2005. "Estimation of the segregation potential of fine-grained soils using the frost heave response of two reference soils." *Can. Geotech. J.* **42** (1): 38–50.

Konrad, J.-M., and N. R. Morgenstern. 1983. "Frost susceptibility of soils in terms of their segregation potential." In *Proc., 4th Int. Conf. on Permafrost*, Fairbanks, AK.

Saarelainen, S. 1992. "Modelling frost heaving and frost penetration in soils at some observation sites in Finland: The SSR model." *VTT Publications 95. Espoo, Finland: Technical Research Centre of Finland. Accessed December 12, 2019.* http://www.vtt.fi/inf/pdf/publications/1992/P95.pdf.

St-Laurent, D. 2006. *Chaussée 2: Logiciel de dimensionnement des chaussées souples. Guide de l'utilisateur.* Montreal: Québec Ministry of Transportation.

CASE STUDY 3

Allen Army Airfield Frost Heave

Kevin Bjella
Ariel Ellison
Jon Maakestad

INTRODUCTION

Typical of northern climates, Interior Alaska is a prime location to observe frost-heave phenomena. There are three conditions that must be met for frost heave to occur: a source of water in some quantity; a soil medium capable of supporting capillary water migration (small pore space) or vertical or lateral wicking of water; and a freezing temperature boundary (freeze front) within close proximity of the water source. During the winter of 2016–2017, evidence of frost heave was detected in the asphalt surface at the Allen Army Airfield (AAAF) at the US Army installation in Fort Greely, Alaska.

The scope of this study was to investigate the mechanisms of frost heave at AAAF by conducting ground-penetrating radar (GPR) surveys, borehole logging, and soil moisture content and grain size distribution analysis at five different sites on the airfield complex. The field work and subsequent analysis for this project was carried out by the Cold Regions Research and Engineering Laboratory (CRREL) in July and August 2017.

AAAF is in Interior Alaska, approximately 8 mi south of the town of Delta Junction. Although there are minimal data concerning the maintenance and operations prior to 2004, it is known that in 2004, runways 10–28 (southeast to northwest) and 1–19 (northeast to southwest) were scheduled for repair and unspecified upgrades for bearing capacity. These upgrades were necessary to handle the heavier payloads required to support US Army Space and Missile Defense Command operations. Between 2004 and 2006, these two runways underwent resurfacing and improvement, including upgrades to lighting and installation of Instrument Landing System equipment (Kaya Associates 2005). During the 2014 construction season, further improvements were made to airfield drainage: infiltration trenches were installed along the west side and southeast corner of the airfield complex.

GEOLOGY

Fort Greely and Delta Junction are between the foothills of the Alaska Range to the south and the Yukon Tanana Uplands to the north. Successive glacial advances have deposited and reworked material in the Delta Junction area by alluvial and glaciofluvial processes. Advances during the Delta glaciation deposited large amounts of coarse moraine material onto the Tanana Valley floor. The more recent Donnelly glaciation did not advance into the valley to the same lateral extent as the previous, coming many thousands of feet short of the previous glaciations but deposited outwash over the Delta-era material. The Delta glacial materials consist of coarse, poorly sorted gravels with cobbles and boulders, as well as sand and silt. The Donnelly glacial material is characterized as massive to well-sorted gravels with cobbles, scattered boulders, and sand (Péwé and Holmes 1964). All deposition in the area is mantled by a layer of windblown silt (loess).

The Donnelly outwash terrace appears to make up the majority of Fort Greely and AAAF's foundation materials. The terrace extends approximately 10 mi south from the airfield complex, bordered by the Delta River to the west and Delta-glaciation-era till deposits to the east. This terrace conforms to a sloping profile, with higher elevation to the south and lower elevation to the north and is completely truncated by Jarvis Creek and the Delta River. The portion of this terrace formation beneath the airfield does not show signs of recent alluvial disturbance. However, owing to its low northern elevation, proximity to active waterways (Jarvis Creek), and observable variability in creek channel location, alluvial reworking of at least part of the airfield foundational material is highly likely. See Figure C3-1 for a layout of surficial geology in the area.

HYDROGEOLOGY

The airfield is just south of the Delta River / Jarvis Creek juncture and is bracketed on three sides by these two water systems, with the Delta River on the west and the Jarvis River on the east and north. Some airfield surfaces are within 600 ft of active braids of Jarvis Creek and within roughly 3,000 ft of the Delta River. Jarvis Creek is a glacially fed and braided waterway and experiences massive flow fluctuations, from a maximum discharge of approximately 883 ft^3/s during months of peak melt to essentially zero flow during the winter, barring warming events (Daly et al. 2011). Most notably, near AAAF, Jarvis Creek becomes influent, losing a substantial amount of flow to the underlying aquifer (Liljedahl et al. 2017). Although less consequential to AAAF, it is important to note that the nearby Delta

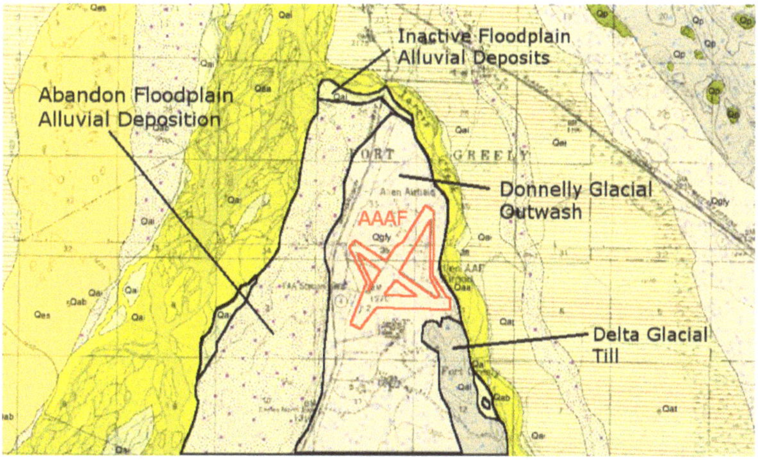

Figure C3-1. Surficial geology of the Allen Army Airfield area.
Source: Reger et al. (2008).

River is glacially fed and braided in nature and exhibits a large degree of channel movement. The Delta River flows year-round, with discharge that ranges seasonally from 450 ft^3/s to 9,000 ft^3/s, depending on glacial runoff (Reger and Péwé 2002). Anecdotal information suggests that the water table is quite deep in the AAAF area. Groundwater approximately 1 mi downstream of the Jarvis/Delta confluence has been observed at 30 to 80 ft in depth (Liljedahl et al. 2017).

INVESTIGATION

Each of the five sites (Figure C3-2) was initially surveyed with a Geophysical Survey Systems SIR-3000 GPR unit and a 400 MHz antenna, providing notable waveform returns down to approximately 15 ft. The team was unable to calibrate the GPR unit to an object of known depth; therefore, absolute depth of subsurface structures could not be precisely determined from the radargrams. However, subsurface structure and layering were clearly identified; and during drilling, the relative depth of a structure (with respect to other subsurface features) could be accurately determined. After gridding each site with GPR, borehole locations were determined by both radargram and physical distress features. Boreholes were drilled with a Geoprobe 7822DT drill rig using direct push. Sampling at AAAF was completed using a MC5 Macro Corer Direct Push Technology (DPT) system that was 2.25 in. in diameter with a 5 ft long sample barrel and clear acrylic liner. The MC5 system enables undisturbed core sample recovery and subsequent visual inspection. Soil samples were collected at three of the five sites and were analyzed for moisture content and soil particle size distribution.

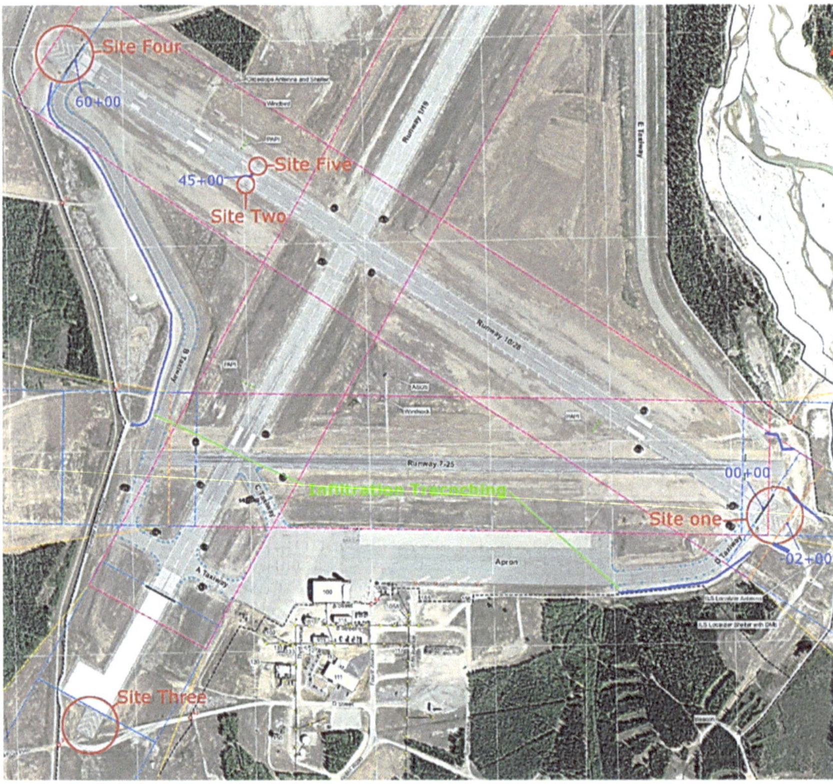

Figure C3-2. Site locations at Allen Army Airfield. Note arbitrary stationing and infiltration trenches. Trenches occur along the west side of the airfield complex, and four others are near Site 1.

Source: Fort Greely, Alaska, Department of Public Works.

Site 1 is on the southeast end of the runway complex and was initially constructed during major renovation of Runway 10-28 during the 2006 construction season. Site 1 is notable for its close proximity to Jarvis Creek (<500 ft) and its profound frost heaving during the 2016–2017 winter. Nineteen GPR transects in a grid pattern were used in the geophysical characterization of the site. Most notably, during the GPR surveys, subsurface structures appeared to vary dramatically along the clearly observable, irregular, and curved distress feature delineated in blue in Figure C3-3. In the radargram, a clearly layered substructure is visible on the northwest (airfield) side of the distress feature (left side of radargram) but is replaced by a more chaotic signal return on the southeast (Jarvis Creek) side (right side of radargram), indicating a change in base material (Figure C3-4).

Based on the GPR results, three boreholes were drilled at Site 1. Borehole 1 (BH-1), on the northwest side of the distress feature, revealed a stable base material of sand and gravel structural fill over a native sand, silt, and gravel mixture observed to be quite dry. BH-2, southeast of the distress feature, revealed a

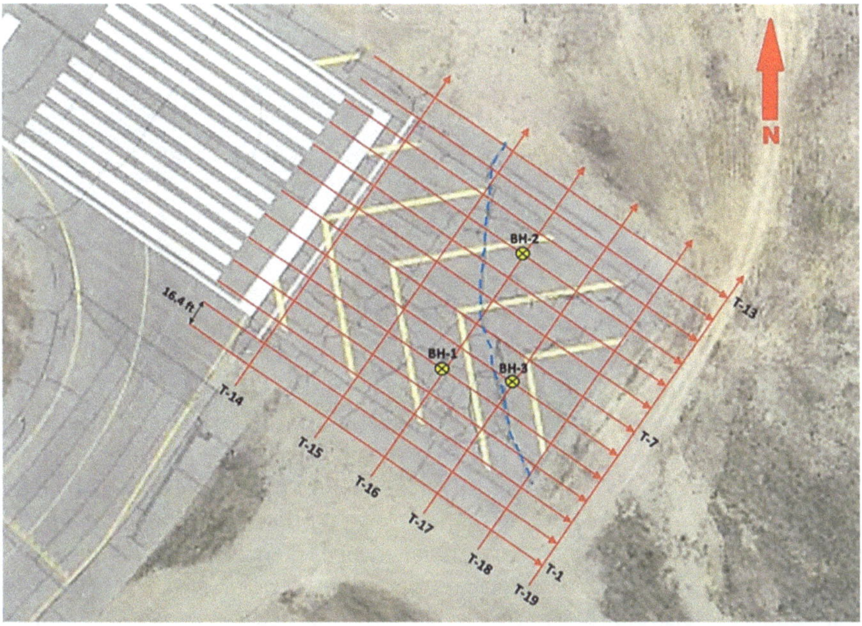

Figure C3-3. Site 1 map outlining ground-penetrating radar transects (red), borehole locations (yellow), and distress feature location and orientation (blue).
Source: Base map provided by Fairbanks North Star Borough GIS Division.

1 in. gravel base course over moist sand, silt, and organic layers approximately 5 ft thick. The silty sand is distinctly layered and overlies a more organic-rich layer of wet silt. The bottom 2.5 in. of the borehole consisted of damp-to-dry sands and gravels. High water content in the midlevel sand-silt-organic layer was consistent with chaotic and bright GPR signals. BH-3 was near the distress feature itself and was chosen to confirm the lithology seen in BH-2. A shallow gravel base layer was found, approximately 4 in. thick, over 4 ft of soft, moist medium sands. Gravelly medium sand layers appeared between 8.5 ft in depth and the bottom of the hole at 10 ft. The sand layers in BH-3 were easily compressed during sampling, indicating poor bearing capacity. The sand and silt layer was again present, but no obvious water sources were located. We assume the fine-grain strata discovered in this area preferentially retains water, which is most probably infiltrating from the cracked asphalt surface. This retained water was theorized to be the source of the recent frost heaving at this location (Figure C3-5).

Site 2 is about midway along the northwest side of Runway 10–28, near the southwest side of the shoulder. As with Site 1, the most recent major paving at Site 2 occurred during the 2006 Runway 10-28 renovation. The location of Site 2 is a concern because of its proximity to the active runway surface. During the 2016–2017 winter, frost-heave-related movement was observed displacing a runway edge marker out of vertical. To reflect the concentrated nature of observed

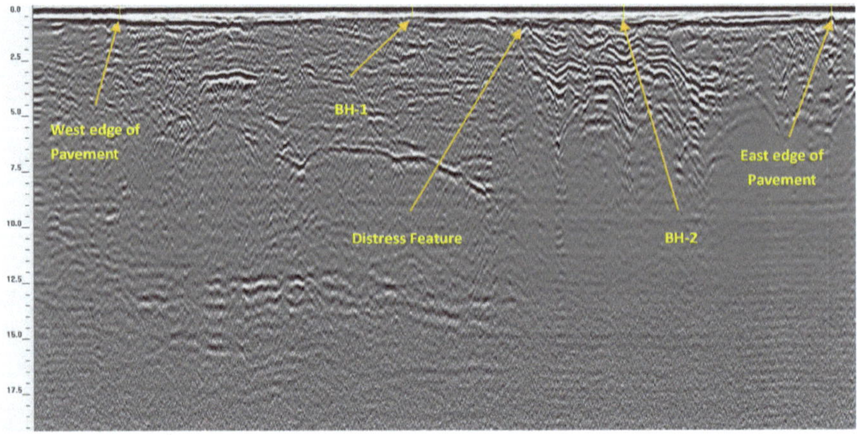

Figure C3-4. Site 1 transect T-16 radargram. Depth is in feet; the image is horizontally compressed. Layers of bright white and black indicate reflected polarity changes. Note the position of the distress feature in relation to boreholes. BH-1 is characteristically sands and gravels typical of structural fill, but BH-2 contains a high prevalence of sands, silts, and organics down to a depth of nearly 8 ft. Again, the distress feature delineates the border between relatively stable subsurface layers to the west and a chaotic, wet subsurface to the east.

Figure C3-5. Frost heave at Site 1, looking down distress feature to the south. Left-side subbase material consists of silts and sands; right side consists of sands and gravels.

Source: Fort Greely, Alaska, Department of Public Works.

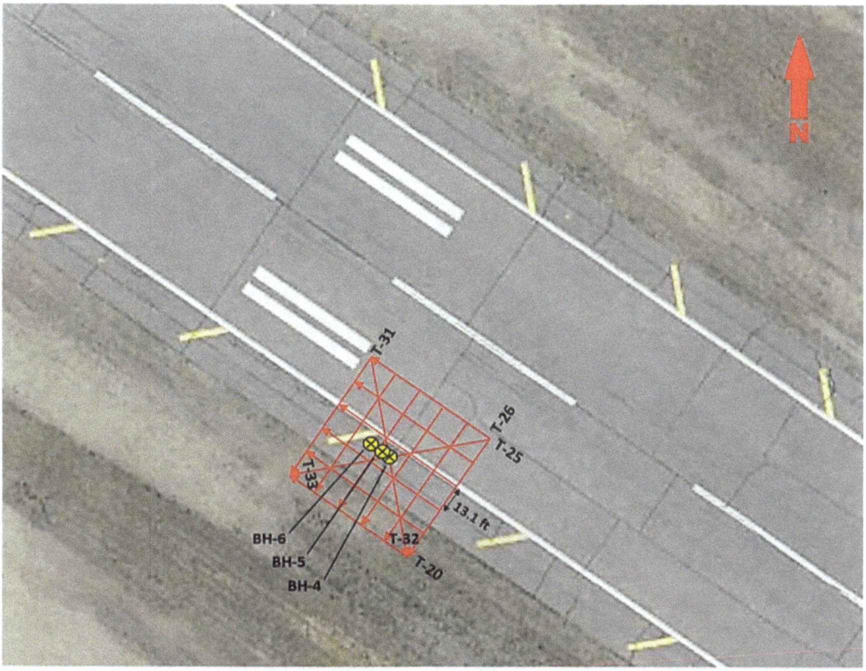

Figure C3-6. Site 2 map outlining ground-penetrating radar transects (red) and borehole locations (yellow).
Source: Base map provided by Fairbanks North Star Borough GIS Division.

frost heave at this site, the GPR survey grid, with 14 transects, was scaled down in size and centered over the runway marker (Figure C3-6). The GPR survey of Site 2 revealed a significant attenuation of radar energy localized in the area of observed frost heave (Figure C3-7). Attenuation of the radar signal may indicate wet subsurface conditions.

In addition to the GPR survey, another three boreholes were drilled at Site 2. BH-4 was placed at the center of an area of the runway shoulder with the most pronounced cracking, where frost heave had been observed. BH-4 revealed a moist layer of oxidized, fine material just under the gravel base course, indicating prolonged exposure to standing water (Figure C3-8). The location and depth of the oxidized layer are similar to the region of attenuated GPR signal. BH-5, 3.4 ft northwest of BH-4, effectively showed the same results as BH-4: oxidization and evidence of prolonged water exposure. BH-6 was 17 ft northwest of BH-4 and showed no signs of oxidization or prolonged water exposure, and subsequently no signs of frost heave during the 2016–2017 winter. The apparent limited range of oxidized fines observed in these three boreholes, coupled with the relatively small area observed to heave, indicated a localized problem.

Site 3 is in the blast zone at the approach end of Runway 1–19, in the southwest corner of the airfield complex. Data concerning the initial construction

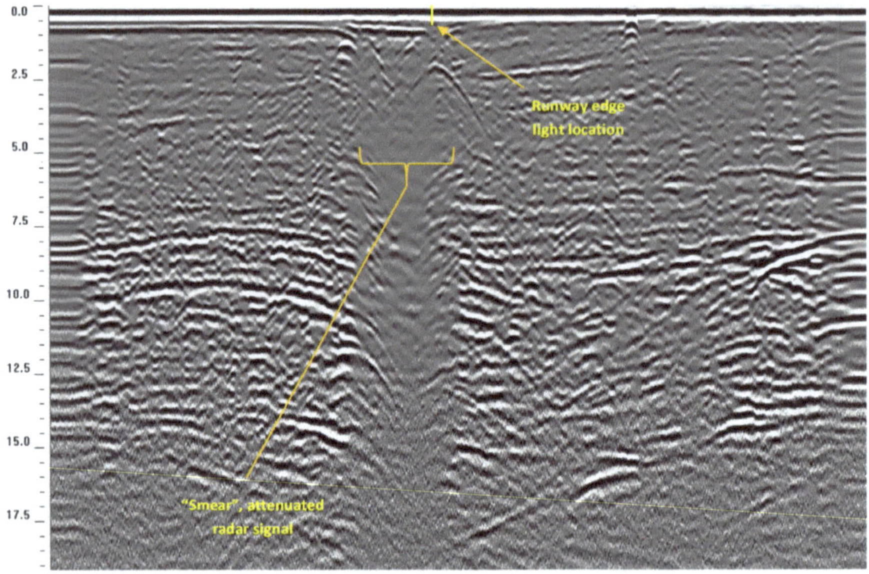

Figure C3-7. Transect T-29 radargram. Depth is in feet; layers of bright white and black indicate reflected polarity changes and, in this case, generally denote changes in moisture content. "Smear" is a region of highly attenuated radar signal, possibly the result of high moisture content just below the asphalt concrete layer. This attenuation casts a "shadow," obscuring detail below it. BH-4 and BH-5 show evidence of prolonged ponding in this same area.

are unavailable. Site 3 was chosen as an area of interest as there was some evidence of frost-heave action, namely large cracking in two transverse bands, visible in Figure C3-9. Whereas these transverse cracks appear similar to the distress feature at Site 1, subsurface conditions remain homogeneous on either side of the feature. The GPR survey at Site 3 consisted of three transects, each aligned with observable surface features (Figure C3-9). All transects displayed a bright band 1 to 2 ft in depth, indicating a strong polarity change and possibly a significant change in soil water content, often indicative of fine-grained strata (Figure C3-10). This band is discontinuous in brightness but discernable to some degree in each radargram. Borehole data suggest that this is a silt layer with higher moisture content than surrounding sands and gravels.

Two boreholes were drilled at Site 3, both near GPR transects. BH-7 was nearest transect T-35 and also near a large transverse crack. The lithology of BH-7 consisted generally of sand and gravel layers but most notably a moist sand silt layer around 3 ft below grade. This moist, fine-grained layer roughly coincides with the polarity-change band visible in the GPR data. The moisture content of Sample 1 at BH-7, taken just above the fine layer, was 18.5% by weight, likely due to water infiltration through nearby cracks. BH-8 was almost in the center of the blast zone and just off the turning point of transect T-36. BH-8 shows a similar

Figure C3-8. Base course is visible on the left; oxidized coloring indicates long-term exposure to water and oxygen. BH-4 was under an area with heavy cracking, both sealed and unsealed. The oxidized layer also contains a relatively high concentration of fines, visible by its texture and separation around the 3 ft mark. BH-5 has similar discoloration but to a lesser extent.

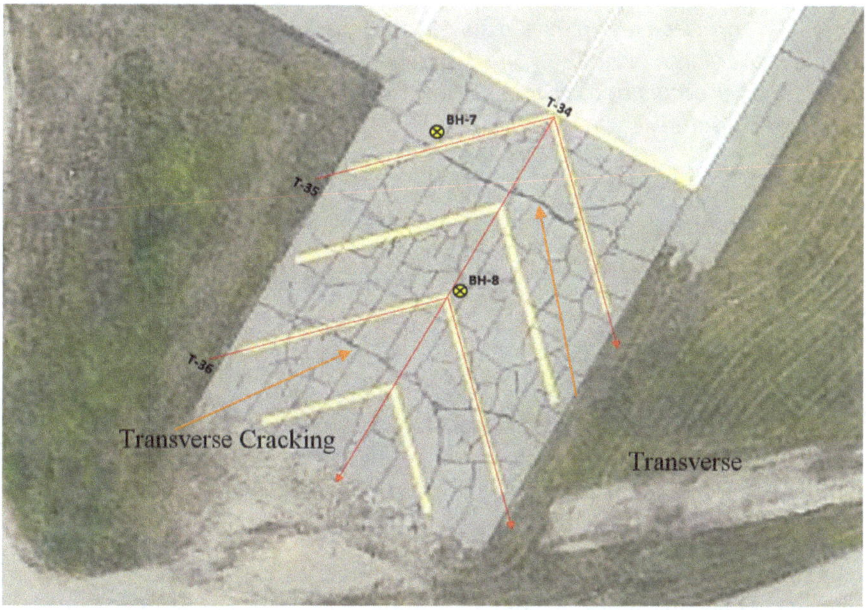

Figure C3-9. Site 3 map outlining ground-penetrating radar transects (red) and borehole locations (yellow). Note the heavier transverse bands of cracking in this area.

Source: Base map provided by Fairbanks North Star Borough GIS Division.

lithology to BH-7. Two samples were from BH-8: Sample 1 in the silty, sandy gravel layer lying above the fine-grained layer at approximately 16 to 26 in. below the surface, and Sample 2 at the upper boundary of the fine-grained layer, approximately 21 in. to 38 in. below the surface (Figure C3-11). The water content in Sample 1 was 2.3% by weight, while water content in Sample 2 at the

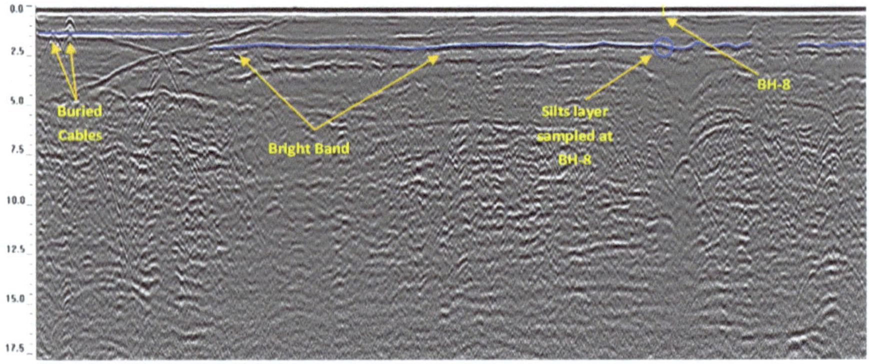

Figure C3-10. Transect T-34 radargram. Depth is in feet; the image is horizontally compressed. Layers of bright white and black indicate reflected polarity changes and, in this case, generally denote changes in moisture content. Delineated in blue is a bright band that was prevalent in nearly all transects at Site 3. Given its continuity and clear return, it is assumed to be a homogeneous layer and is prevalent over large portions of Site 3. Bore logs of borehole BH-8 indicate that a moist silt layer occurs at the same depth, and it is assumed to be responsible for this radar signature.

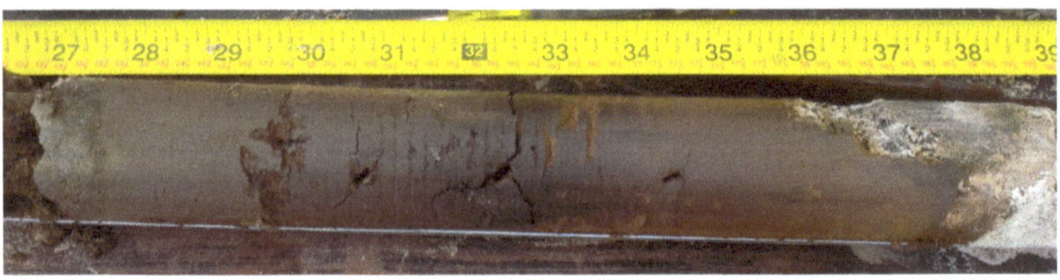

Figure C3-11. Example of the silt layer present in BH-8. This material occurs at roughly the same depth as the bright band noted in Figure C3-10.

upper boundary of the fine layer was 12.6% (Figure C3-16). From the collected lithological data and moisture content, Sample 1 in BH-7 and Sample 2 in BH-8 both indicate that the fine-grained layer contained much more water and is capable of aggravating frost heave conditions.

Site 4 is on the northwest end of Runway 10-28 and was most recently reconstructed during the runway renovations of 2006. Site 4 appeared much more stable than all previous sites; the area was well crowned and showed less evidence of frost-heave action and no major cracking bands (Figure C3-12). Seventeen GPR transects were conducted over this site location, most focused over the blast zone. Transects T-49 through T-54 were located in such a way as to explore possible differences in surface or subsurface construction methods between active and inactive runway surfaces (Figure C3-12). "Inactive" refers to runway surfaces not

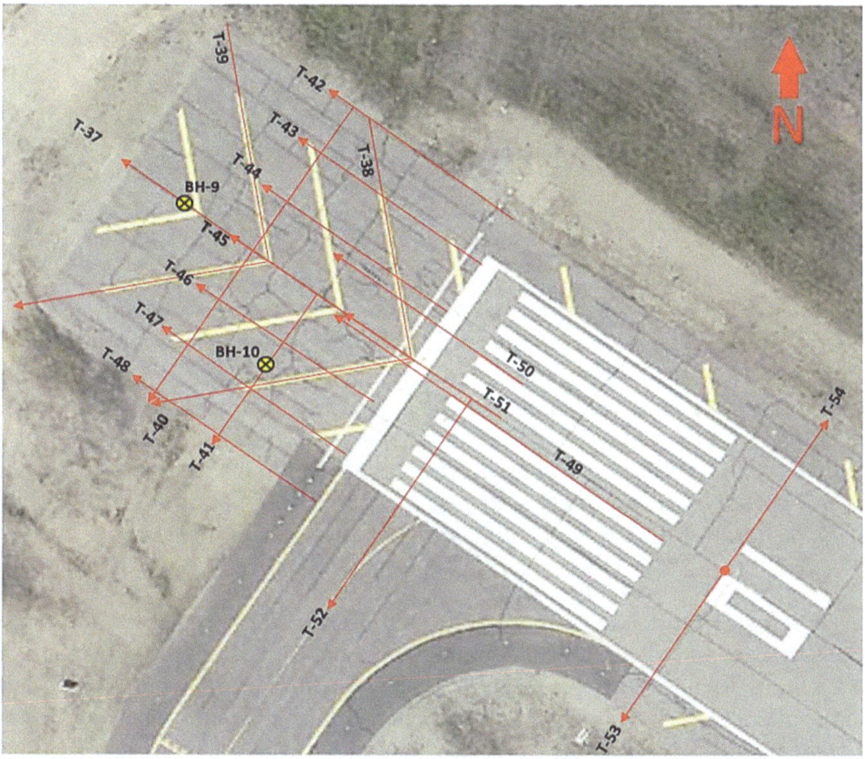

Figure C3-12. Site 4 map outlining ground-penetrating radar transects (red) and borehole locations (yellow). Note the smaller amount of cracking compared to Site 3.
Source: Base map provided by Fairbanks North Star Borough GIS Division.

subjected to regular aircraft wheel traffic. The most notable anomaly visible in the GPR surveys of this site was a bright band in the radargram, possibly indicating high moisture content, consistent with fine-grained materials (Figure C3-13). The depth of the bright band varied in relation to the keel of the runway; the band was slightly deeper near the center than near the outer edges. This difference is probably due to the depth of base course required to create the crown of the blast zone.

On completion of the GPR survey, two boreholes were drilled at Site 4. BH-9 was along the peak of the crown and along GPR transect T-37. The lithology of BH-9 was consistent with other boreholes and can be characterized as mostly sand and gravel layers. However, like Site 3, a definite silty layer was present at a depth corresponding to the bright band visible in GPR returns. Samples from the silty layer in BH-9 contained a high percentage of fine material (58% < 200 microns) and a moisture content of 22.8% by weight. BH-10 was on an old frost blister and intersected GPR transect T-41, which was conducted after drilling. This borehole

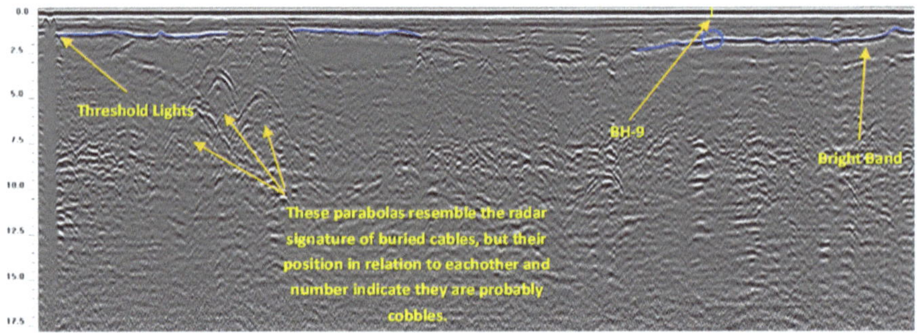

Figure C3-13. Transect T-37 radargram. Depth is in feet; the image is horizontally compressed. Layers of bright white and black indicate reflected polarity changes and, in this case, generally denote change in moisture content. The bright band is visible and nearly continuous in T-37. The discontinuity in radargram signature could be a result of variance in moisture content instead of composition. This silt layer is probably present under large portions of the site but not as readily visible to radar because of its low moisture variance relative to its surroundings.

revealed a wet/moist silty layer just below a thin base course of gravels and sand. The silty layer was about 50% fine-grained material, with a high moisture content of 22.1% by weight. A sample taken just below the fines layer was primarily sand and contained much less water (3.8% by weight). This difference is taken to mean that the fines layer is retaining water. See Figure C3-16 for gradation and moisture content of samples from Site 4 boreholes.

The final site, Site 5, is about midway on the northwest side of Runway 10-28, across from Site 2, on the northeast side of the shoulder (Figure C3-14). As with Site 2, the most recent paving at this location occurred during the 2006 Runway 10-28 renovation. Two transects were conducted, both transverse to the spine of Runway 10-28 and extending beyond the boundaries of the shoulder on each side. The purpose of these longer transects was to examine possible transitions in subsurface lithology relative to the runway surface, shoulder surface, and unpaved areas beyond the shoulders. Transect T-55 runs between two transverse cracks, extending from the spine of the runway onto shoulder surfaces. Transect T-56 runs over a section of runway without substantial surface cracking (Figure C3-14).

Four boreholes were drilled at Site 5, three following transect T-55 and one on transect T-5 (Figure C3-15). BH-11 was north of the paved shoulder area and consisted of largely native sands and gravels but no definite silt layer. BH-12 was in the middle of the shoulder at Site 5. The lithology of BH-12 was consistent with that of the other shoulder boreholes: base course fill over native sands and gravels. No major silt deposits were present in BH-12. BH-13 was just inside the runway edge markings, on the runway surface, and in the center of a sealed crack. BH-13 returned a 12 in. thick asphalt concrete (AC) layer over silty sand and gravel with no defined base course. Sand and gravel layers directly below the AC layer contained 4.2% water by weight and provided relatively easy drilling. BH-14 was

Figure C3-14. Site 5 map outlining ground-penetrating radar transects (red) and borehole locations (yellow).
Source: Base map provided by Fairbanks North Star Borough GIS Division.

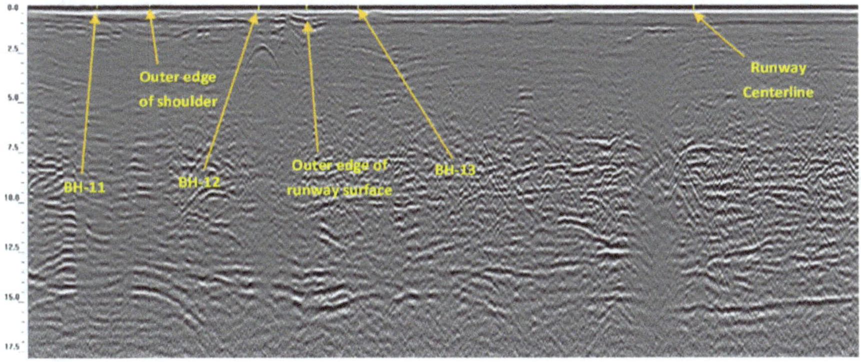

Figure C3-15. Transect T-55 radargram. Depth is in feet; the image is horizontally compressed. Layers of bright white and black indicate reflected polarity changes and, in this case, generally denote changes in moisture content. The runway surface seems to overlay homogeneous materials. Boreholes 11, 12, and 13 showed no signs of the silt layer present at Sites 3 and 4.

35 ft west of BH-13, just inside the runway edge markings and in a section of runway without prominent cracking. BH-14 provided notable drilling resistance from the surface to the bottom of the hole, indicating good compaction and solid foundation material. The lithology of BH-14 consisted of 17 in. of AC over a thin (4 in.) layer of base course and silty sand. Below the AC and base course were layers of silt, sand, and gravel. Samples taken to a similar depth as BH-13 contained 3.3% water by weight, less than BH-13, possibly because of the lack of surface cracking and water infiltration.

SUMMARY OF FINDINGS

The subsurface material at Site 1 transitions from relatively stable subbase gravels on the northwest side of the distress feature (Figure C3-3) to fine silts with sand on the southeast side. The transition is abrupt and easily discernable, perhaps the result of an old slough or former braid of Jarvis Creek. The rapid channel movements of Jarvis Creek, noted from year to year, support this hypothesis. However, the age and extent of this old slough bed are indeterminate at this time. The fine-grained soils and wet conditions on the southeast side of the distress feature are probably the mechanism and cause of seasonal frost heave, while the northwest side, with its more stable subbase material, appears less vulnerable.

The Site 2 frost heave is likely the result of subsurface ponding, as shown by the oxidized layer of material in BH-4 and BH-5. This ponding appears to be localized to the immediate frost heave area, as no oxidization was noted at BH-6, 17 ft away. The source and path of the moisture at Site 2 are still unclear at this time, but flow may be occurring through unsealed cracks in the runway or shoulder surfaces. Without forensic excavation of the area, the water's origin and path cannot be satisfactorily explained. It is likely that the water in this area is causing the relocation of fine material (piping), and forensic excavation is necessary to better identify the issue.

Sites 3 and 4 display similar characteristics, with discernable silt layers just below the pavement. While frost-heave action appears to be more prevalent at Site 3 than at Site 4, both have conditions conducive to frost-heave formation. Unsealed cracks in the AC layer at both sites are avenues for precipitation to make its way into the substructure. The layer of fine material (silt layer) can hamper drainage of the sand, silt, and gravel layers above it and retain water longer than its surroundings. This source of water, and the small pore space of fine-grained material, greatly increase the likelihood of frost heave.

Site 5, while not displaying notable frost-heave movement or an obvious layer of fine materials in lithology of boreholes or GPR radiogram, does show evidence of water infiltration. Specifically, at BH-13, the base course and subbase were drilled with little resistance, relative to the base course and subbase material at BH-14. The likely explanation of this difference in density is water infiltration through surface cracks at BH-13 but not at BH-14.

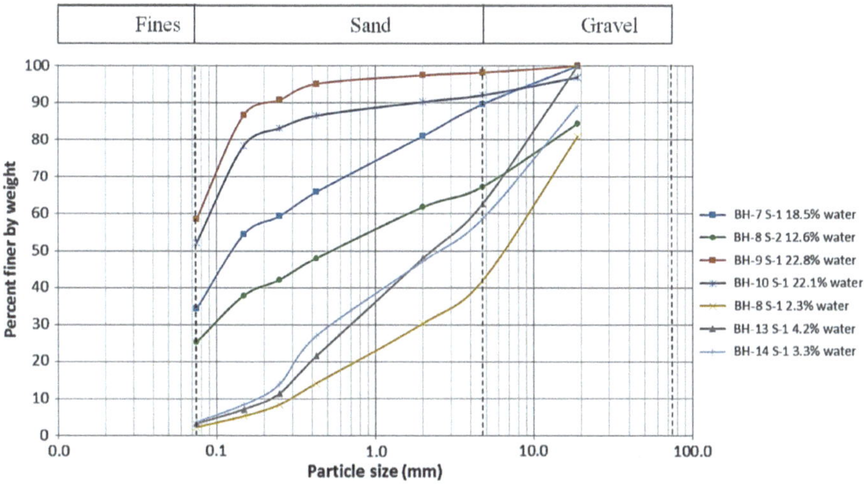

Figure C3-16. Gradation curves of samples. Note the correlation of a high percentage of fine material and a high water content. BH-13 and BH-14 are samples from a stable area with no evidence of frost heaving. Moisture content is low in both samples, even though BH-13 was directly under a crack and showed other evidence of water-related destabilization (i.e., low drilling resistance in the sample area). Samples in BH-8 are adjacent to one another; Sample 2 is a fines-rich layer, and Sample 1 is just above it, yet water content varies by a factor of six.

Frost heave occurs when three main conditions are met: water in some quantity, soil capable of supporting capillary motion or wicking, and a freeze front close to the water source. CRREL's investigation was conducted in summer with the primary focus to identify fine soils capable of wicking moisture from water sources. In all cases, it is assumed that water originates at the surface and permeates through cracks into subsurface structures, as opposed to migration up through sand and gravel from the groundwater table 30 ft below. These five sites represent varying conditions, and, as such, varying frost-heave severity. Figure C3-16 partially illustrates the relationship between percentage of fine material and moisture content. Thick fine layers and high moisture content resulted in significant frost heave at Site 1. A restricted fines layer and ready source of moisture at Site 2 resulted in definite frost heave motion, but less than at Site 1. Sites 3 and 4 had different degrees of frost heave, although they both contained a fines layer over large portions of the site. Site 4 is crowned, while Site 3 is not; this may have resulted in less water infiltration at Site 4 and thus less frost-heave-related motion. Finally, Site 5 exhibited evidence of water infiltration; but with no fines layer to retain or transport it, frost heave did not occur.

References

Daly, S. F., J. E. Zufelt, P. Fitzgerald, A. Gelvin, and S. Newman. 2011. *Aufeis formation in Jarvis Creek and flood mitigation*. ERDC/CRREL TR-11-14. Hanover, NH: USACE,

Engineer Research and Development Center, Cold Regions Research and Engineering Laboratory.

Kaya Associates. 2005. *US army space and missile defense command (SMDC), Fort Greely, Alaska (FGA), SMDC mission summary development plan for FGA*. Huntsville, AL: Kaya Associates.

Liljedahl, A. K., A. Gädeke, S. O'Neel, T. A. Gatesman, and T. A. Douglas. 2017. "Glacierized headwater streams as aquifer recharge corridors, subarctic Alaska." *Geophys. Res. Lett.* **44** (13): 6876–6885.

Péwé, T. L., and G. W. Holmes. 1964. *Text and description of geologic units for the Mt. Hayes D-4 Quadrangle, Alaska*. Reston, VA: USGS.

Reger, R. D., and T. L. Péwé. 2002. *Geologic map of the Big Delta A-4 Quadrangle, Alaska: Report of investigation RI 2002-2*. Fairbanks, AK: State of Alaska, Dept. of Natural Resources, Division of Geological and Geophysical Surveys.

Reger, R. D., D. S. P. Stevens, and D. N. Solie. 2008. *Surficial geology of the Alaska Highway Corridor, Delta Junction to Dot Lake, Alaska*. Fairbanks, AK: State of Alaska, Dept. of Natural Resources, Division of Geological and Geophysical Surveys.

Index

Page numbers followed by *f* and *t* indicate figures and tables.